DR. LEE SPETNER

THE EVOLUTION REVOLUTION

WHY THINKING PEOPLE ARE RETHINKING THE THEORY OF EVOLUTION

THE EVOLUTION REVOLUTION
Why thinking people are *rethinking* the theory of evolution

© 2014 Dr. Lee M. Spetner

All rights reserved. No part of this publication may be
translated, reproduced, stored in a retrieval system,
or transmitted in any form or by any means, electronic,
mechanical, photocopying, recording, or otherwise,
without permission in writing from the publisher.

ISBN: 978-1-60763-155-2

Also by this author:
Not By Chance! Shattering the Modern Theory of Evolution

Editor: Nachum Shapiro
Proofreader: Hadassa Goldsmith
Cover design and layout: Justine Elliott
Internal design and layout: Nachum Shapiro

The Judaica Press, Inc.
123 Ditmas Avenue / Brooklyn, NY 11218
718-972-6200 / 800-972-6201
info@judaicapress.com
www.judaicapress.com

Manufactured in the United States of America

*To Julie, of blessed memory,
with whom I shared
a wonderful lifetime*

Contents

Preface 7

1. Evolution and Information 13

2. The Nonrandom Evolutionary Hypothesis 41

3. Rapid Evolution Is Happening Today! 67

4. The False Arguments for Evolution 85

5. Epilogue 129

Acknowledgments 146

References 147

Index 161

Preface

IN 1963, AS I WAS LISTENING TO A LECTURE ON EVOLUTION BY Bentley Glass at Johns Hopkins University, I wondered how all the information in those plants and animals could have been built up. I had been teaching statistical communication theory and information theory at the time, which is probably why this problem occurred to me. I asked Bentley about it a few days after the lecture, but he couldn't help me. He had never thought about my problem. My curiosity about this issue continued and eventually led to my publishing a few papers on the subject, in *The Journal of Theoretical Biology* (which was a rather new journal at that time), in *Nature*, and in *Transactions on Information Theory of the IEEE*. But my ideas on the subject had not yet matured.

I continued to feel that there was something not right about the theory of evolution. I later read Richard Dawkins's book *The Blind Watchmaker*, and I felt it didn't make sense. Random errors in copying the DNA and natural selection were supposed to account for the evolution of all life from some simple primitive cell. I could not understand how random mutations and natural selection could account for the information buildup in what is called Common Descent. Could those DNA copying errors really bring new information into living organisms? I decided I had to set straight a lot of things about evolution that Dawkins wrote. As a result, I wrote my book *Not By Chance!*, first published in 1996 in Israel and then in 1997 in the United States.

Neo-Darwinian Theory (NDT) is counterintuitive, and is acknowledged as such even by its supporters. All present-day life is assumed to have evolved from some primitive cell, and that cell was supposed to have formed itself from simple chemicals. Nobody seems to know how that cell came to be, but almost all biologists think they understand fairly well how evolution proceeded from that cell to all the life we see today.

There appears to be a vast amount of information contained in trees, fish, elephants, and people. Where did this information come from? It is said to have come from random mutations and natural selection. How can that work? Natural selection is supposed to be the magic that makes evolution happen, but all natural selection does is eliminate the less adaptive organisms and allow the more adaptive ones to survive and proliferate. Where do those more adaptive ones come from? Apparently, that's what random mutations are supposed to accomplish.

So the information buildup required by Common Descent can come only from random mutations. That means that the buildup of information is a matter of chance. At each step of the evolutionary process, a mutation has to have occurred that grants the organism an advantage. The big question is: Is that reasonable? To see if it is, some people (including me) have made calculations of the probability of mutations building information.

We really don't have all the data we need to make this calculation. But even if we make some conservative assumptions and give the benefit of all doubts to the Darwinian side, such calculations demonstrate that Common Descent is not reasonable (Spetner 1997). The Darwinists, however, do not accept these calculations as conclusive — they suggest alternative scenarios that might make the probabilities larger.

In his book *Darwin's Black Box*, Michael Behe addressed the unreasonableness of Darwinian evolution. He described some biological systems as what he called "irreducibly complex." By that he meant that these systems are composed of several critical components in such a way that the system cannot work unless all those components are in place. He then argued that the system could not evolve one small part at a time, because natural selection could not work on less than the whole system. Here, too, the Darwinians countered by suggesting scenarios in which natural selection *might* work, but again, the Darwinian scenarios are purely hypothetical.

Because the Darwinians can invent scenarios to address any challenge to their theory, they are not convinced by attempts to show that neo-Darwinian evolution cannot work. Therefore, I have concluded that it would be more productive to challenge *them* to show that it *could* work — challenge them to do more than just offer vague scenarios of how their theory *might* work, but to show by calculation that the probability of it working is reasonably high. This is a challenge they must meet to establish their theory on a scientific basis. They have never met this challenge and they cannot. They cannot show that the events they claim to have produced Common Descent have a high enough probability to justify their claim. Their inability to establish the theory of Common Descent means that Common Descent is not an established theory. This is one of the main points of this book.

I cannot overemphasize the importance of probability calculations. NDT is not like Newton's theory of mechanics, whose equations describe the motion of a physical body under a force. Nor is it like Maxwell's theory of electromagnetism, whose equations describe the effects of electric and magnetic fields on electric charges. These theories are checked against experiment by solving those equations. NDT describes evolution as the result of random mutations that may or may not yield an adaptive phenotype. These are chance events. The theory can be checked only by calculating the probabilities of the required events to see if they are reasonably large. The theory has not been shown to have passed this test and is therefore not a valid theory. Whatever evidence is given for Common Descent is circumstantial. Circumstantial evidence cannot stand alone. It needs to have a theory tying the evidence to the conclusion. But instead of a theory, imaginary scenarios are offered to suggest how evolution *might* work. No calculations of probabilities are made.

The fossil evidence for evolution consists of examples of extinct organisms that are assembled into an order showing a transition from one form to another. That evidence, however, is a long way from a convincing case for evolution. Because these organisms are extinct, there is no way to show what is actually claimed — namely, that the transitions were effected by a long sequence of small changes. The theory claims, for example, that some ape-like creature, because of a mutation in a sex cell, gave birth to a slightly more human form. Then one of that creature's descendants did the same, and so on. This chain continued for a few million years until finally a full human was born. This sequence has not been observed, yet that is the claim of Common Descent. Real evidence of Common Descent is nonexistent. I deal with this in Chapter 4.

In contrast to the lack of evidence for Common Descent, there is an abundance of evidence for some kind of evolution. Populations are indeed observed to change. For decades, Darwinists have held Common Descent is just an extension of these observed evolutionary changes. The most famous of these is the evolution of antibiotic resistance in microorganisms, but there are many others. I give examples in Chapter 3.

In my study of the literature during my research for my earlier book, I was led to realize that plants and animals appear to have a built-in capability to change in response to an environmental input. These changes are heritable, and are passed on to subsequent generations. They are often (and perhaps always) adaptive and are indeed evolution. But for at least two reasons they are not the kind of evolution NDT requires for Common Descent. For one thing, these changes happen much more rapidly than evolution is supposed to occur — much more rapidly than could be explained by random mutations and natural selection. For another, they add no information to the genome, which they would have to do if they are to contribute to Common Descent.

In my earlier book, I outlined an hypothesis of this limited kind of evolution. I called it the NREH, or *Nonrandom Evolutionary Hypothesis*. The hypothesis is that organisms have a built-in mechanism that enables some environmental stimuli to trigger genetic changes that can lead to adaptive phenotypic changes, and that these changes can be heritable. Both the mechanism for these genetic changes and the environmental triggering mechanism are built into the organism. I do not speculate on the origin of these mechanisms, and there is no evidence of how they might have originated, nor is there a theory to indicate how it could have evolved.

I offered the NREH as a speculation based on observations of rapid adaptive evolution and observations of spontaneous genetic rearrangements that are often adaptive. How these genetic changes could be stimulated by the environment was not clear to me at that time, but it nevertheless seemed to be happening. Now, fifteen years later, I have found lots of evidence in the literature supporting the NREH. There is now evidence of epigenetic mechanisms that make it work. I discuss this in Chapter 2.

The NREH can not only shed light on evolution, but it also holds the promise of solving other outstanding problems. I discuss this in Chapter 5. Most important, the NREH does a better job of accounting for evolutionary data, such as the fossil record and biogeographic diversity, than does the random-mutation/natural-selection theory of neo-Darwinism. I discuss this in Chapter 4.

My critique of neo-Darwinian theory is thus twofold. Firstly, a crucial requirement of the theory — that the probabilities of the random mutations are high enough to make the theory reasonable — has never been established. Secondly, the usual evidence given for the theory does not support the theory. I offer another explanation of that evidence that does not share the deficiencies of that theory and conclude that neo-Darwinian theory must be retired.

Lee M. Spetner
Jerusalem, Israel
June 2014

Chapter 1

Evolution and Information

As I write these words, several of my great-grandchildren are running through the house, chasing each other, laughing and yelling. I have observed three generations born, develop, and mature. I am constantly amazed at how a tiny fertilized egg develops into a sentient human being; how he[1] becomes aware of his surroundings and even of himself. That little brain feeds a mind that thinks, learns, and even generalizes. I remember when my oldest son was barely three years old and ran to me excitedly saying,

"I know how much is three threes!"

"Oh," I said, interested in his discovery, "how much?"

"Nine," he said.

"That's right!" I said, with some pride in my first offspring.

"Do you want to know how I know?" he continued.

Now I was really intrigued. "Yes," I said. I surely wanted to know whatever story he had to tell about how he discovered that.

"Come with me; I'll show you!"

---

[1] In English it has been traditional to use the masculine pronoun even when the antecedent could be feminine or masculine. In the past few decades, however, many authors, to avoid an imagined offence to the distaff side, use the slashed combination he/she in such cases. I think this alternative is clumsy and detracts from the thought flow. Other authors go all the way to the other side and use she instead of the traditional he. I don't see this as an improvement. For simplicity, and in keeping with tradition, I shall use the masculine pronoun with no offence intended.

He ran up the stairs to his room, with me after him.

"Look," he said, pointing to the linoleum on the floor that had repetitive patterns of three bears and three pigs. "Three bears, three bears, and three bears: 1, 2, 3 — 4, 5, 6 — 7, 8, 9! Three bears three times are nine bears! Now look over here. Three pigs, three pigs, and three pigs: 1, 2, 3 — 4, 5, 6 — 7, 8, 9! Three pigs three times are nine pigs!"

Then came his conclusion: "Three threes are nine!"

I wonder if he wasn't quite sure if "three threes are nine" might be a property only of bears, until he saw it also worked on pigs, and only then did he make the intellectual jump. I don't know what went on in that little one's mind, but he certainly did make a correct generalization.

I don't know what the mind is, but that mind, whose coordinate brain did not exist barely three years earlier, had made what I thought was a profound generalization. You can't program a computer to make such a generalization without having it make many very silly ones as well. Darwinists are trying to convince us that all life, including the mind, has developed from inert chemicals through the ordinary workings of physics and chemistry. I find it difficult to believe that life, and in particular the human mind, spontaneously developed and were the result only of the known forces of nature. And I am not the only one who has this difficulty. This difficulty is apparently widespread, as Darwinists have recognized. The natural origin of life is so counterintuitive that Francis Crick was compelled to write:

> Biologists must constantly keep in mind that what they see was not designed, but rather evolved. (Crick 1988, 138)

Richard Dawkins wrote in a similar vein:

> The feature of living matter that most demands explanation is that it is almost unimaginably complicated in directions that convey a powerful illusion of deliberate design. (Dawkins 2003, 79)

Dawkins is correct; it certainly does demand an explanation.

Life surely appears to have been designed, but many claim the appearance is deceptive, and life was not designed at all but evolved through events that are essentially random. This claim is counterintuitive, and to establish such a claim demands evidence, good evidence. In actuality, there really is no good evidence for it.

Darwinists use two different definitions of evolution. The first definition is a modification over time in the heritable characteristics of a population, which they call a change in its gene pool. The second definition is that all living organisms descended, with modification, from an ultimate, very primitive, ancestor — usually taken to be a single cell. Most of them hold that there was only one ultimate ancestor and all life descended from a single source. The process is called Common Descent. In this definition, evolution relies on random heritable changes and natural selection.

The two definitions are very different. The first, which is just population change, can be observed in the field and demonstrated in the laboratory. In later chapters I shall describe how these changes occur and give many examples. The second, Common Descent, is not demonstrable at all. Examples of population change do not prove Common Descent.

Common Descent is a key component of an agenda advocating a natural origin of life. The effort to demonstrate the possibility of such a natural origin is usually divided into two parts: (1) *abiogenesis*, the origin of a simple life form from naturally occurring chemicals, and (2) the evolution of all life from that single simple beginning. It turns out, however, there is no good evidence for either of these two parts. The evidence usually offered for the second part, based on fossil, biogeographical, and other data (e.g., Coyne 2009, Dawkins 2010, Rogers 2011) is seriously flawed, as I shall demonstrate.

I am critiquing evolution not under its definition as simply population change, of which there are countless examples that we see happening all around us. Rather, I am critiquing the theory of Common

Descent — the alleged "grand sweep of evolution" — wherein all of the complex life of today is supposed to have evolved from some very primitive cell according to the second definition above. I will show that there is no acceptable evidence for Common Descent and that the evolution we actually observe cannot be extended to produce Common Descent *no matter how long a period of time it has to potentially work.*

Evolution is not the only counterintuitive theory in science. Quantum mechanics (QM) is also counterintuitive, but there is a lot of good evidence for it. Indeed, QM was only reluctantly accepted because it successfully accounted for data that classical physics was at a loss to explain. Classical physics could not, for example, explain black-body radiation, which could be properly explained only by assuming that electromagnetic radiation was quantized in discrete packets having energy proportional to the frequency. QM has turned out to be an extremely successful theory in that it accounts for observed data with amazing precision. Although I have not used QM in my professional life, I did teach an informal course in the subject to my colleagues in the Cosmic-Ray Group at MIT when I was a graduate student there more than sixty-five years ago. Since then, QM has been shown to have even more counterintuitive theoretical consequences, which have been borne out by experiment.

The acceptance of QM was *data*-driven because it accounted so well for the data, and it was accepted for that reason only. The acceptance of the theory of evolution, on the other hand, seems to have been *agenda*-driven because Darwin proposed it despite its failure to account for the data. Indeed, Darwin was worried that the fossil record, as he knew it, appeared to contradict his theory.[2] Moreover, the thought of the evolution of the eye gave Darwin a "cold shudder, which," he said,

---

2 Darwinists have since managed to explain away the lack of transitional fossils, but that does not change the fact that to Darwin the fossil record was an obstacle to his theory that had to be explained away. He proposed his theory in spite of the counterintuitive aspect of Common Descent and in spite of the data. I shall deal with the fossil record in Chapter 4.

"my reason tells me I ought to conquer" (Darwin 1860). Nevertheless, he pursued his theory, apparently driven by his agenda.

The second part of the program of finding a theory of the natural origin and development of life is a theory of the development from an alleged simple, primitive beginning to all the abundant forms of life we see today. This is the subject of Darwin's work and he is generally given credit for having solved it by his suggestion of random variation and natural selection. Since his suggestion relies on a preexisting, self-replicating organism, there must also be a theory for how the first self-replicating organism arose in a natural way, and this is Part 1 of the program: finding a theory that will account for abiogenesis, the natural origin of a simple form of life from nonliving matter. The life generated according to such a theory can be as simple as one can conceive — some simple cell perhaps — but it must, at the very least, be capable of metabolizing, i.e., exploiting energy from the environment, and of self-reproduction to allow for the Darwinian process in the second part of the program.

In Darwin's day, there was no clear understanding of what a living organism was. There was no clear understanding of what distinguished living from nonliving matter. I remember as recently as seventy-five years ago there was a belief, even among some scientists, that the content of a cell was inherently different from inanimate matter in that it had some mysterious *vital* component that was not contained in inanimate matter, and only through this mysterious component was it alive. I even had a biology teacher in high school in 1940 who called herself a *vitalist* and held living matter was alive because of some "vital" component it contained. As strange as *vitalism* may seem today, it turns out it was not completely off the mark. Today we understand that there is such a "vital" component of a cell, but it is no longer so mysterious. It turns out that this "vital" component is necessary (although it may not be sufficient[3]) to

---

[3] We won't know whether or not it is sufficient until we are able to synthesize a living cell from inert chemicals.

make the cell "alive." That component is *information*. Today we routinely input information into our computers, although that does not make them alive. There is, however, no inanimate object that contains and uses information that wasn't put there by an intelligent being.

The world of the inanimate is made up of two entities, matter and energy. Life requires a third component — *information*. In the latter half of the twentieth century, it became apparent that an important distinction between living and nonliving matter is the information in the living organism, and the capability of the organism to use that information to perform its functions. The information of life is presently thought to reside to a large extent in the molecules of deoxyribonucleic acid (DNA) that are the constituents of the genome (the set of genes and their controls) in every living cell, and is therefore called *genetic information*.[4]

The DNA molecule is a chain of small molecules, called nucleotides, of which there are four types, denoted by the letters A, T, C, and G.[5] The DNA molecule plays the role of a message made up of a sequence of these nucleotides, which play the part of letters in the message. The message is thus written in an alphabet of four letters. As we understand today, the information in the genome of an organism plays an essential role in the development of the form of the organism and how it functions. The way the genetic information controls the working of the organism is thought to be somewhat the way a computer program controls the working of a computer, although it appears to be much more complex. When an organism reproduces itself, it copies the genetic information and bestows the copy on its progeny.[6] In sexual

---

[4] The popular fascination with DNA has created the impression that all, or at least most, of the information in a cell lies in the DNA. There is, however, a lot of information in a cell that is not in the DNA. Without that other cellular information, the DNA wouldn't do anything.

[5] For a concise description of the DNA molecule see my book *Not By Chance!* You can also learn about DNA from almost any high school biology text.

[6] Of course, the entire cell replicates, so all the other cellular information also replicates.

reproduction, the genetic information of both parents is copied and partitioned into the progeny.

Charles Darwin is credited with showing how to solve Part 2 of the program — Common Descent — in 1859 with the publication of his *Origin of Species*, but he understood the need for Part 1 — abiogenesis — as well. He envisioned a primitive cell forming spontaneously from the chemicals available in a "small warm pond" (Darwin 1871). Thomas Huxley (1869) tried his hand at solving abiogenesis, as did Henry Bastian (1870). But the attempt to prove abiogenesis is considered to have had its real beginning in 1924 in the work of the Russian biochemist, Alexander I. Oparin, who studied how a membrane that could enclose some molecules could form, isolating them from the outside world.

The complexity of living things, which we are only beginning to understand today, was not appreciated in those days. A cell was thought then to be a bag of chemicals that somehow fulfilled the functions of life. The complex and precise workings of even a simple cell would not even begin to be revealed for another thirty years. So Oparin believed that if a membrane would form that could serve as a bag to enclose the right chemicals, it would become a cell and then somehow the functions of life would start working. Oparin's intent was that a primitive cell formed in this way could absorb nutrients from the outside and thereby perhaps grow and reproduce itself (Oparin 1953 [originally published in Russian, 1924]). Exactly how the contents of that membrane could become a cell that could grow and reproduce was left to future study. Part 1 picked up momentum with the work of Stanley Miller (1953), who showed a few amino acids could be produced from inert chemicals in a reducing atmosphere in the presence of electric sparks. But the appearance of a few amino acids is a long way from the appearance of life. Neither of these studies addressed the crucial question of the origin of the *information* for life.

The problem of how life could have come from non-life has been intensively engaged by many brilliant scientists for almost a century, but the results have been consistently disappointing. Although theories have been proposed for how some of the small biologically significant molecules could form spontaneously, no progress has been made in discovering how they could become joined together to generate the necessary information. There is no theory that can account for the spontaneous appearance of the information of life that exists in even the most simple and primitive living cell.

Leslie Orgel was a biochemist who, until his recent death in 2007, was one of the leading scientists working on abiogenesis. He wrote:

> There is no agreement on the extent to which metabolism could develop independently of a genetic material. In my opinion, there is no basis in known chemistry for the belief that long sequences of reactions can organize spontaneously — and every reason to believe that they cannot. The problem of achieving sufficient specificity, whether in aqueous solution or on the surface of a mineral, is so severe that the chance of closing a cycle of reactions as complex as the reverse citric-acid cycle, for example, is negligible. The same, I believe, is true for simpler cycles involving small molecules that might be relevant to the origins of life and also for peptide-based cycles. (Orgel 1998, 494–495)

There is no theory today that satisfactorily accounts for the origin of a living organism from inert chemicals. For decades there have been optimistic declarations that success in proving abiogenesis is just around the corner. But each time we have turned the corner it hasn't been there.[7]

Part 2 (Common Descent), on the other hand, is confidently declared to have been accomplished. Darwinists have claimed success in accounting for the natural development of life from a simple beginning. But to establish Common Descent they are obligated to present a theory of how the information contained in the many complex life

---

7 Logically, the whole discussion should end here, because as Darwin himself recognized, without abiogenesis Common Descent cannot begin. But for those who think Common Descent is a successful theory and that its success might portend the eventual discovery of a theory of abiogenesis, I shall explain the fallacy of Common Descent.

forms of today has been built up from a primitive cell that was so simple it could have arisen from inert matter. That primitive cell would have to have been very simple indeed, especially in view of the lack of success in showing that abiogenesis is possible at all. In what follows, I shall examine their claims of success with Common Descent.

Since nearly the beginning of the twentieth century, Darwinists have proclaimed both the *theory* of evolution and the *fact* of evolution. The *fact* of evolution, they say, is that all present forms of life arose from ancestral forms different from themselves. Darwinists claim, for example, that "birds arose from nonbirds and humans from nonhumans" (Lewontin 1981) and that "humans evolved from ape-like ancestors" (Gould 1981). The *theory* of evolution, on the other hand, is the mechanism offered to explain how the "fact" came about. In reality, though, their *theory* is not a theory, nor is their *fact* a fact.

What can they mean by a "fact"? Is it indeed a *fact* that some ape-like creature gave birth to a slightly more human-like offspring (male? female? both?) and that one of its descendants gave birth to a still more human-like offspring, and so on until one of its descendants gave birth to what we would consider a fully-human offspring? A fossil find is a fact. But a hypothetical chain of events that has not been observed cannot be reasonably called a *fact*.

How did that unobserved chain of alleged events ever come to be called a *fact*? These alleged events are not like the fact that an apple falls to the ground when dropped. It is not like the fact that a magnetic field causes a movement of a wire carrying an electric current. If it is any kind of fact at all, it is of the historical kind. An historical fact is normally considered to be one for which there is documentary evidence or other hard evidence of what is claimed. There is no evidence of ape-like creatures giving birth to offspring that are more human-like in a sequence of steps that eventually leads to the birth of a fully-human being. That is not a "fact" that can qualify as historical or as anything else.

There is no evidence for such a sequence of births. The best that can be said is that there is some meager circumstantial evidence in the fossil record. But for circumstantial evidence to support a claim, there must be a theory that connects the claim to the evidence. Random events (genetic mutations) are said to have effected this sequence of births of increasingly human-like apes. In this case, there must be a probability calculation showing that what they are suggesting is reasonably likely to have occurred. No analysis has been offered to show this. The existence of fossil "intermediates" alone is not compelling evidence that one gave birth to the other. I shall elaborate on this point in Chapter 4.

Molecular evidence of a similarity in the proteins or DNA in humans and present-day apes is even less compelling as evidence that the two had a common ancestor from whom they arose gradually, one birth after another. One researcher wrote:

> The same genetic code, the same DNA, the same amino acids and the genetic message that unites all organisms, independent of morphology, proves that the theory of evolution is as well established as any in science. (Yockey 2005, 181)

Proves? The best that can be said is that *if* Common Descent were true, then all organisms would exhibit the same DNA and amino acids. The logic, however, does not run the other way. The same backward logic has been applied to the fossil evidence.

For the theory to connect the circumstantial evidence with the conclusion of Common Descent, the theory must account for the buildup of the biological information and complexity present in all life today. It must account, in particular, for the buildup from the information in the alleged primitive cell to that, for example, in the human brain, which has been called the most complex object in the universe (Steen 2007). It is incumbent on Darwinists to show how the theory of evolution can do this, and they have not done so. They have instead offered only scenarios of how, in some vague way, some evolutionary changes *might* have happened. These scenarios fall far short of what is needed for a theory required to establish the counterintuitive claim of Common Descent.

At the heart of the process described by the modern theory of evolution are random changes in the heritable information (they speak mainly of the DNA) of the organism. Darwinists often say evolution is *not* random. They claim that although the theory posits random mutations, the process of natural selection acts as a filter that picks out the "better" of these random changes, thus producing a sequence of more and more adaptive forms from a sea of random alternatives. Such a procedure, they allege, can build the complexity we see in life today by selecting from randomly generated novelty. But allegations alone cannot establish a theory.

According to the theory, the novelty lies only in the random genetic changes. The randomness of mutations is fundamental to neo-Darwinian theory, the modern refinement of Charles Darwin's work.

According to neo-Darwinian theory, the information of life is principally the genetic information residing in the DNA. Evolutionary theory must account for the increase in the genetic information, which is claimed to have developed from the information in the alleged first primitive cell to the vast amount of information found in life today. The proposed mechanism consists of a two-component process: random genetic changes, followed by natural selection. Why must the theory make essential use of random genetic changes? Randomness has been chosen because there is no known *deterministic* process by which the information could be built up. A deterministic process would, for example, be one where the chemistry of nucleotides causes them to combine preferentially into the information-bearing chains of DNA found in living organisms rather than into chains of nonsense. Since no such chemical affinity for information building has been observed, or is known even to be theoretically possible, random genetic changes followed by natural selection has been chosen as the method of building information and complexity.

The laws of chance lie at the heart of the process of natural selection not only because the mutations are governed by the laws of chance, but also because the process of natural selection must include chance deaths of individual organisms, whether they are adaptive or not. Because evolutionary theory has randomness at its core, the primary requirement for the theory is to show that the probability of evolution actually building up the required information in this manner is reasonably large. Wolfgang Pauli, who won the Nobel Prize in Physics in 1945, and who is considered one of the greatest physicists of the twentieth century, objected to evolutionary theory because the very crucial probability calculations had not been addressed. About the theory of evolution he wrote:

> Als Physiker möchte ich hier das kritische Bedenken geltend machen, dass dieses Modell bisher durch keine positive Wahrscheinlichkeitsbetrachtung gestützt ist. Eine solche müsste in einem Vergleich der aus dem Modell folgenden theoretischen *Zeitskala* der Evolution mit ihrer empirischen Zeitskala bestehen: *es müsste gezeigt werden, dass auf Grund des angenommenen Modelles de facto vorhandenes Zweckmässiges eine genügende Chance hatte, innerhalb der empirisch bekannten Zeit zu entstehen. Eine solche Betrachtung wird jedoch nirgends versucht.*
>
> [As a physicist, I would like to express here some critical misgivings that this model has so far not been subjected to any positive probability analysis. Such an analysis should consist of a comparison of the *time scale* of the theoretical evolutionary model with the empirical time scale: *it would have to be shown, on the basis of the assumed model, that the* de-facto *existing adaptations have a sufficient chance of occurring within the empirically known time. Such an analysis has, however, never been attempted.*] (Pauli 1954) (My translation. Emphasis in the original.)

More recently, Thomas Nagel, a professor of philosophy at New York University, has expressed the same misgivings:

> ... for a long time I have found the materialist account of how we and our fellow organisms came to exist hard to believe, including the standard version of how the evolutionary process works ... What is lacking, to my knowledge, is a credible argument that the story has a non-negligible probability of being true. (Nagel 2012)

Without an analysis of the probabilities to substantiate the theory, *there is no theory of evolution.* Without a calculation showing the probabilities to be *significant*, evolution is no more than a collection of stories.

Newton's theory of gravitation was a deterministic theory; he could calculate that his inverse-square law of gravitation would determine that the moon, for example, would travel in an elliptical orbit around the earth. He could verify his theory by showing that the orbits followed by the heavenly bodies were the very ones he calculated. The correspondence between his calculations and observations was evidence for the correctness of his theory.

The theory of evolution, on the other hand, is not deterministic and is not verifiable by showing that the observed complexity is built up in a deterministic way. It is instead based on chance phenomena and chance phenomena are properly examined by calculating probabilities. One must verify the theory by showing the probability is reasonable that today's observed complexity can be built up by random mutations and natural selection. The Darwinists have not shown by calculation that such buildup is reasonably probable. Whatever calculations have been made show the probabilities to be forbiddingly small (Eden 1967, Spetner 1997). Instead of making probability calculations, Darwinists tell stories about how such complexity *could have* been built, but stories do not establish a scientific theory. Darwinists will tell you they have made calculations, but those calculations are of things like how a mutation spreads through a population. They have not addressed the issue of the probability of an *adaptive* mutation. They have not shown it to be reasonably likely that an adaptive mutation will *appear*. And they have certainly not shown how it will happen over and over again, each time improving adaptivity, until great complexity has been built up. These are the calculations that must be made to establish the theory. Darwinists have not made them.

I shall show you in this book that there is no compelling evidence for the so-called *fact* of evolution. In the past, when I have shown what is wrong with the *fact* of evolution, I have been criticized for not proposing a replacement naturalistic theory that overcomes my objections and is yet able to account for the origin and development of life. "How, then, did life originate?" they ask. Of course, the critics insist that the theory I propose lie within Science as we know it today, and I cannot do this. But I argue that if a theory cannot account for the facts, it has to be discarded even if there is no replacement. If the best you can get is no good, you shouldn't accept it.

To say we must accept the neo-Darwinian theory (NDT) because it is the best we have is not a valid argument in its favor. The rejection of a theory does not necessarily require that there be a replacement. The philosophical consequences of accepting the NDT and the "fact" of evolution are too important to consider anything less than a proper theory.

* * *

In 1982, 26-year-old Alton Logan was arrested and charged with the murder of a security guard during a robbery at a McDonald's on the south side of Chicago. Alton was charged with murder and brought to trial. During the trial he was identified as the murderer by an eyewitness.

The buzzing in the courtroom fell silent as the judge called the court to order. The jury filed in and took their seats in the jury box. The judge turned to the jury and asked: "Has the jury reached a verdict?"

"We have, your honor."

"What is your verdict?"

"We have found the defendant guilty of murder in the first degree."

When the court reconvened, the judge had to pronounce sentencing based on the verdict of the jury.

"Will the defendant please rise," said the judge.

The young man rose to hear the judge announce, "Following the finding of the jury, I hereby sentence you to life imprisonment without the possibility of parole."

A shriek went up from the young man's mother. "He's innocent!" she cried. "He didn't kill anyone!"

The young man who, as it was later found, was indeed innocent, was sent to spend the rest of his life in prison. How could this gross error have happened? His brother and mother had testified that he was sleeping at home at the time of the murder. The murder weapon was not found in his possession, but in the possession of someone else who had a criminal record and whom he did not even know. There was no physical evidence linking him to the murder. The only testimony the prosecution brought against him was an eyewitness who identified his photograph as the murderer.

Alton Logan spent the next twenty-six years of his life in Joliet prison for a crime he did not commit. The case of the prosecution was weak — eyewitness testimony is known to be unreliable [Wells and Olson 2003]. But the district attorney had to have a conviction and pressed his case with the jury even though he knew it was weak and his suspect may indeed have had nothing to do with the murder. The case was weak but no better suspect was found, so this innocent man had to be sacrificed to give the police and the prosecution a conviction they needed. Despite the many holes in the prosecution's case, it was the best they had, so this suspect had to be the one to be convicted. Would it not have been better if no one were convicted than to have convicted the wrong man? Why did there have to be a conviction if it wasn't the right one?

It turned out that, after he had served twenty-six years of his life sentence, two attorneys came forward with an affidavit they had signed at the time of the trial attesting that Andrew Wilson, the real killer, had confessed to them but denied them permission to reveal it until after his death. They had locked the affidavit in a strongbox at the time of the conviction and brought it out only after the real killer died. (Alton was released from prison, and in January 2013 the State of Illinois awarded him $10.2 million dollars as compensation for the years he spent in prison.)

* * *

It is much the same with the theory of evolution. There are advocates of evolution who acknowledge that there are serious weaknesses in the theory but when challenged reply, "Do you have a better theory of how life arose and developed?" Creation is, of course, not allowed. One may ask why Creation should be ruled out, but that question will not get a hearing today in the life-science community. If one cannot offer a better theory of how life could have come about through known physical laws, then according to their rules the theory of random mutation and natural selection wins. They therefore conclude, by default, that life arose by random mutations and natural selection. But if the theory is inadequate, then why must its conclusion be accepted — a conclusion fraught with nihilistic consequences? Darwinists contend evolution (by which they mean Common Descent) is as well established as any scientific theory. But this contention is not true, as I shall demonstrate.

Just as the main problem with demonstrating abiogenesis is to show how information got into the alleged first simple primitive cell, the main problem with Common Descent is to show how information in the alleged primitive cell got built up to the level of complexity and information contained in today's living organisms. When the Darwinists invoke the fossil record or draw up a list of vestigial organs as arguments for evolution, they are not addressing this critical issue. Although the alleged buildup of information in the past is not observable, to support Common Descent one must at least offer a theoretical mechanism as to how such a buildup could have occurred. Moreover, one should also present, insofar as possible, observations of such buildup occurring today according to that mechanism.

If there were a theoretical mechanism that could indeed account for the buildup of the information of life, and if, in addition, one could somehow demonstrate abiogenesis, that would offer some support for those who choose to accept a natural origin for life and deny the necessity of a Creator. In the present state of the theory of evolution, however, the claim that evolutionary theory provides a scientific support for atheism is false.

What measure of success has been achieved with Common Descent?

The Darwinists claim success by being able to explain many aspects of living organisms in terms of evolution and Common Descent, but those explanations rely on an unlikely assumption. They assume that whenever there is a need for an adaptive change in a population, an appropriate mutation is available. When the need for a change arises, for example, they assume that the right mutation — occurring at random and independent of the need for adaptivity — is available for natural selection to act upon. However, there is neither theoretical nor observational support for such an assumption. They also claim several supports for Common Descent, such as the fossil record and biogeographical data. In Chapter 4, I shall examine these as well as other claims.

Note that since the Darwinists must assume that the beneficial mutations are very small, indeed the smallest possible, then for significant evolution to occur, there must have been long chains of evolutionary steps in each of which an adaptive mutation had occurred, followed by natural selection. This assumption can be justified only if it is backed by a calculation of the probability of beneficial random mutations occurring and being successively more adaptive so natural selection can operate at each step. But as I have noted above, they have not made that calculation.

A recent attempt was made to address an aspect of this problem where the authors concluded that the mean length of time for evolution to produce a string of adaptive mutations was proportional to the logarithm of the length of the string (Wilf and Ewens 2010). Had their result been correct, it might have offered some support for Common Descent. Unfortunately for the theory of evolution, however, their analysis has been shown to be flawed. For one thing, they did not address the issue of the probability of occurrence of an adaptive mutation. For another, they had a confused idea of evolution. Their conclusion is therefore invalidated (Spetner 2011).

Computer simulations of the evolutionary process have been developed with the claim they show that evolution works, and in particular with the claim that random mutations and natural selection can build up information. I have previously dealt with Richard Dawkins's program (Dawkins 1986) and have shown why it fails to represent Darwinian evolution (Spetner 1997). More recently, Tom Schneider has written a program (Schneider 2000) called **ev**, claiming:

> ... contrary to probabilistic arguments by Spetner (1997, 1964), the **ev** program clearly demonstrates that biological information, measured in the strict Shannon sense, can rapidly appear in genetic control systems subjected to replication, mutation and selection.

Contrary to what Schneider claimed here, his program demonstrates no such thing. His simulation is flawed in that it fails to represent important features of evolution. The high mutation rate he requires for information buildup would, in real life, drive a species to extinction. While his high mutation rate can indeed generate information, it will at the same time destroy essential functional elements of the genome. Schneider's program has no provision for simulating essential portions of the genome that would be vulnerable to the high mutation rate, and his simulation therefore fails to capture the essence of Darwinian evolution.

Under special conditions, a sufficiently high mutation rate, together with selection, can indeed build information. An example is the generation of antibodies in the vertebrate immune system.[8] However, these mutations are of the *somatic* type, which means they do not occur in the germ line, or gametes, and are therefore not available for evolution. They occur only in the genome of the B lymphocytes, and only in a special restricted portion of that genome, and only when the body is under pathogen attack. The mutation rate of this section of the DNA has been found in mice to be as high as 10^{-3} per base pair per generation (McKean et al. 1984), which is about a million times the average mutation rate in the human genome. A high mutation rate can yield not only all possible single-nucleotide changes, but it can yield doublets and even triplets to present for selection, making it possible to gain information

---

8 I am indebted to Ed Max for providing me with this example.

in the presence of selection. However, if this high mutation rate were to occur in the gametes instead of being restricted to a small portion in the B lymphocytes, the species would die out. So, while Schneider's computer simulation program makes an interesting game, it does not show that Darwinian evolution can work.

The Darwinist position is that all of that information was built up though long sequences of random mutations and natural selection. If it were true that innumerable random adaptive mutations occurred in the past, then the same process should be going on now as well and at least some random adaptive mutations should be observable today. Although we cannot observe the evolutionary process acting over millions of years, we should at least be able to observe some small portions of this process. To what extent have any such portions been observed?

The Darwinists offer examples of evolution in action, but none of them, it turns out, can lead to Common Descent. These include the famous example of the evolution of melanism in peppered moths (Kettlewell 1955). They also cite the evolution of antibiotic resistance. I will examine the details of the Darwinists' best examples in Chapter 4 and show they are incapable of being extended to yield Common Descent. Darwinists say, "If these examples of evolution can occur in only a few decades or even years, imagine what extensive evolution could occur in millions of years!" But they can say this only by ignoring and obscuring the important details of the processes that are at work in these examples.

* * *

If an individual animal in a population obtains a heritable advantage, enabling it to give birth to more progeny than the others, then its own progeny will, in turn, inherit the same advantage. They will themselves tend to have more progeny than those without this advantage, and the numbers of their descendants will thus tend to increase exponentially over time. This is the process called natural selection. There is, therefore, a chance that the genome carrying such an advantage can, after many generations, become a majority of the population.

Eventually it could take over the population. If this happens, the population becomes poised for the appearance of another adaptive genetic change that will again grant its possessor a selective advantage over the rest of the population (NAS 2008). Long chains of such adaptive events form the model of the neo-Darwinian theory of evolution.

Natural selection is said to bring direction to the evolutionary process. Random genetic changes, which include mutations and recombinations, create genetic variety in a population, and the process of natural selection is supposed to determine which new genotype, if any, will be emphasized in the population. Natural selection is assumed to bring direction to evolution by eliminating from the population individuals and their genomes that are less suitable in the current environment. Natural selection has no creative ability. It tends to enhance the numbers of the more successful (adaptive) organisms in a population by culling the less successful (less adaptive), but it creates no novelty by itself. In evolutionary theory, the task of adding novelty as genetic information is the burden of the random genetic changes.

The genetic variety within a population is the raw material of evolution. Natural selection can choose only from what is available. The creative side of evolution lies with genetic changes, which generate this variation, and not with natural selection. Novelty appears only through genetic changes that occur in the gametes (sex cells), which are copied and passed on to the next generation. A potentially adaptive genetic change may have occurred in the gametes of the immediate preceding generation, or it may have survived after a population change that occurred in an earlier generation and did not yet disappear. According to evolutionary theory, all genomic variation had to arise through random changes in the genome. The theory also holds that these changes occur by chance and their occurrence is independent of their effect on the individual (the phenotype[9]) that arises from them. According

---

9 The phenotype of an organism is its set of physical characteristics, including its anatomy, physiology, biochemistry, and behavior.

to the theory, all new information and new complexity that appear in living organisms in the process of evolution can originate only from random genetic changes.

There is a popular fallacy that gene duplication brings novelty to the genome. But duplicating a gene doesn't add information to the genome, just as buying a second copy of a newspaper doesn't give you any more information than you had in the first one. The novelty brought through gene duplication comes only through the genetic changes that may occur in the duplicated gene. The supposed advantage of gene duplication is the possibility that the original gene can continue to function while the duplicated one is free to mutate (Zhang 2003). Ultimately, however, novelty comes only from mutation.

Darwinists tacitly assume the available genetic variation to be unlimited; it is either already in the population, having arisen from earlier genetic changes, or it is available from current mutations. They assume that whenever a population needs an evolutionary change, an appropriate mutation will be available, natural selection will act on it, and the population will change accordingly. However, those who make this assumption are obligated to justify it. It is not sufficient simply to cite mutation rates, which can produce raw variation. To justify the assumption, one must show there are enough mutations that would be *adaptive* in the current environment. Darwinists invoke this assumption in the stories they offer to explain the appearance of new features that have never before existed. Examples of this reasoning abound in the literature. A typical one is the following from Speth et al. (2007):

> "… pathogens have *evolved* a myriad of virulence factors that allow them to manipulate host cellular pathways in order to gain entry into, multiply and move within, and eventually exit the host for a new infection cycle." (My emphasis)

They could have more easily written "pathogens have a myriad of…" instead of "pathogens have *evolved* a myriad of …" They don't know that the virulence factors evolved any more than they know they were created.

The life-sciences literature is replete with such examples of an assumption of evolution. But it isn't only the life scientists. Scientists in other disciplines assume evolutionary biologists know what they are talking about, and they have tacitly accepted their assumption of evolution. For example, Donald Canfield, a geologist and professor at the University of Southern Denmark, and his collaborators wrote (Canfield et al. 2010):

> "Atmospheric reactions and slow geological processes controlled Earth's earliest nitrogen cycle, and by ~2.7 billion years ago, a linked suite of microbial processes *evolved* to form the modern nitrogen cycle with robust natural feedbacks and controls. (My emphasis)

How do they know the microbial processes of today have evolved to form a nitrogen cycle a few billion years ago? The fact is they don't; they simply believe what the Darwinists tell them. Barbara Wright, a professor of biology at the University of Montana, wrote the following (Wright 2000):

> "Attenuation regulation [of amino-acid biosynthesis] is an impressive example of the remarkable mechanisms *that have evolved* to ensure the conservation of precious reserves and the derepression and activation of only those systems essential for survival ..." (p. 2994) (My emphasis)
>
> "Presumably, feedback mechanisms existing today *evolved* in the past to prevent unnecessary and wasteful metabolic activities by coordinating these activities with the presence or absence of nutrients in the environment." (p. 2994) (My emphasis. At least here she said "presumably.")
>
> "Mechanisms *must have evolved* in starving cells to stimulate metabolic changes and mutations that facilitate adaptation to new circumstances." (p. 2995) (My emphasis)

Wright does not know that these mechanisms have evolved, yet she calls them "impressive" examples of evolution. In any other field of science, unsupported statements of this sort would be rejected by a scientific journal. But there seem to be different rules for evolution. One of the most egregious of such examples was written by the late François Jacob, a brilliant biologist who received the Nobel Prize in Medicine in 1965. He wrote:

"What is clear, however, is that, like the rest of our body, our brain is a product of natural selection, that is, of differential reproductions accumulated over millions of years under the pressure of various environmental conditions." (Jacob 1977, 1166)

Is it really clear that natural selection produced the human brain? It's certainly not clear to me, and I don't believe it is at all clear to François Jacob how random mutations and natural selection could have produced the human brain — the most complex object in the universe. He has, of course, not explained how this could have happened.

An editorial in the prestigious scientific journal *Nature* was just as bad in making the following outlandish statement:

"… the idea that human minds are the product of evolution is not atheistic theology. It is unassailable fact." (*Nature* 2007)

This is propaganda, not science.

There is no justification for the assumption that if a capability was needed by an organism it would have evolved. The validity of the assumption requires that there be a reasonable probability that an adaptive mutation will occur or is already in the population. This has never been shown. Moreover, the theory requires not just one such mutation, but long sequences of them. Unless the Darwinists can show their assumption is reasonable, they do not have a viable mechanism, and without such a mechanism there is no support for a theory of Common Descent.

* * *

After the discovery of the structure of DNA (Watson and Crick 1953) and its method of replication (Meselsohn and Stahl 1958), the way mutations caused evolution was thought to be understood (Anfinsen 1959). The understanding of mutations was that they were copying-errors in the replication process, which came to be known as "point mutations." The probability of a point mutation in the human gamete is about 7×10^{-9}, or seven in a billion, per nucleotide per replication (Drake 1999). How many potential point mutations would be of selective benefit in any particular environment? The truth is, we don't know of even one that could qualify as a contributor to Common Descent.[10]

I shall not here compute the probabilities of getting the right mutations and of evolving. I have done that in my previous book, *Not By Chance!* (Spetner 1997, Ch. 4).[11] The onus of calculating the probabilities really falls on the Darwinists, who conjecture that long sequences of mutation and natural selection have occurred to yield Common Descent.

Darwinists contend that evolution is not random because natural selection filters the mutations, and they seem to think it must then lead to building information. But filtering alone does not create information. The only possible source of information is the mutations. Any appearance of new functions, structures, or behavior requires the buildup of heritable information. Darwinists are obligated to show how information could have been built up. They are advocating a theory that is based on the random events of adaptive mutations. To establish their theory, they must give evidence of information being built by up random mutations. They have not done this, and most likely they cannot.[12] They must also show that the occurrence of the long sequences

---

10 There are some point mutations in the genome that are known to be of selective value, but these are all of the type that loses information, such as those that disable a repressor molecule permitting a gene to command an unlimited rate of synthesis of its encoded protein. Such mutations may be of selective value under particular circumstances, but they do not add information to the genome and cannot contribute to Common Descent. This will be discussed further in Chapter 4.

11 Tom Schneider criticized my calculation of the probability and posted his criticism on his website: http://schneider.ncifcrf.gov/paper/ev/AND-multiplication-error.html. However, it is his critique that is in fact wrong. My response appears on http://www.uncommondescent.com/evolution/lee-spetner-responds-briefly-to-tom-schneider/. He posted his response to my critique on his website, which indicated that he did not understand my reply. I posted my reply to his reply here: http://www.uncommon-descent.com/intelligent-design/lee-spetner-responds-to-tom-schneider/. This is the same Tom Schneider who published the flawed evolution simulation about which he made false claims and which I discussed above.

12 Some Darwinists have pointed to an example of a sequence of three random mutations that successively improve adaptivity. This is the same example I have previously described in detail (Spetner, 1997) and I have shown that this sequence cannot be extended and cannot contribute to Common Descent. I shall discuss this example briefly in Chapter 4.

of random events of evolution they are conjecturing is even potentially possible.

How much information has to be built up in Common Descent? Well, we don't really know with any assurance just how much information is in the mammalian genome, but we do know the human genome has about three billion nucleotides. So the genome of the alleged first primitive cell, in the process of Common Descent, would have to grow to about three billion nucleotides. That "first" genome could not have had much more than about a million nucleotides. The smallest genome of a free-living organism we know of today is that of a strain of the bacterial family Methylophilaceae, which has 1.3 million nucleotides (Giovannoni et al. 2008). The genome of the alleged primitive cell would not be expected to be larger than this. Hence, in the process of Common Descent, about three billion nucleotides had to be added to the genome, whether by gene duplication or by any other conceivable process. Furthermore, through point mutations, recombination, or any other genetic changes on these duplicated genes, some portion (perhaps most of it) had to have a specific arrangement of nucleotides. This arrangement (or arrangements) had to be built up one nucleotide at a time by a sequence of random substitutions. Moreover, each of these tiny component genetic changes in the sequence had to be adaptive. To substantiate their theory, Darwinists are obligated to show there can even *be* a sequence of changes like this. It is not clear at all there can be. And if they should be able to show there can be such a sequence, they have to show that the probability of it happening is reasonable. They have not done so, and most likely they cannot.

Furthermore, natural selection is a fickle force. Organisms can die or be killed accidentally regardless of how potentially adaptive their genetic makeup is. I have shown previously (Spetner 1997, Ch. 3 and notes) that chance is also strongly involved in natural selection itself. Animals with an adaptive mutation may perish in spite of their adaptive advantage. A single individual can perish through accidents or by being attacked by a predator before it can reproduce. Therefore, even if a mutation occurs that is potentially adaptive, it is highly probable that it will nevertheless eventually disappear from the population (Fisher 1958).

The French naturalist Jean Baptiste de Lamarck proposed a theory of evolution that eventually became discredited. Many high school biology textbooks use the following fictitious scenario to characterize Lamarck's theory and contrast it with Darwin's. Primitive giraffes were short-necked, they say, like many other mammals of their time. They fed by browsing from tree branches. Once the low-lying branches had been stripped by the hungry giraffes, the remaining branches were a little too high for them to reach. They were therefore forced to stretch their necks to reach the higher branches. The stretching made their necks a bit longer, and their offspring inherited the longer necks. The next generation inherited not only the longer neck but also the desire and ability to stretch their necks further. After many generations, with the continued neck-stretching, the later generations of giraffes had much longer necks than their early ancestors. This fantastic and unprovable Lamarckian explanation of the origin of the giraffe's long neck was eventually rejected and replaced by Darwin's explanation. It turns out, though, that Darwin's explanation is no less fantastic and no less unprovable.

Almost all point mutations are recessive (Orr 1991), which means that a mutation in a gene will not have an effect on the phenotype unless it appears in both copies of the gene.[13] Moreover, almost all mutations are harmful, and about eleven percent of mutations in the wild, as estimated from Drosophila, are actually lethal (Dobzhansky 1938). A lethal mutation in a gene is one that will lead to the death of an organism that has it in both its copies of the gene. The mammalian genome has about three billion nucleotides. A change in any one of them during a replication is a mutation. What fraction of those three billion

---

13 Sexually reproducing organisms have their chromosomes in pairs, where each one has the full complement of genes.

possible mutations would be an advantage to the animal in any particular environment? This crucial question has never been answered.

* * *

Point mutations are not the only genetic changes that can occur. In the past half century a host of more complicated genetic rearrangements have been discovered. As Darwinists have begun to recognize that random point mutations cannot account for Common Descent, some are beginning to suggest that more complex genetic rearrangements, which are known to occur and which are usually nonrandom, might do the trick.

As stated previously, the human genome is comprised of about three billion DNA base pairs. The number of genes (coding for protein and RNA) in the genome is currently estimated to be about 20,000. If we take a typical gene to have about 1,000 base pairs, then we can estimate that the genes constitute less than 1% of the genome. What, then, is the role of the other 99% of the genome?

About forty years ago the late Susumu Ohno, who was well known in the field of molecular evolution, suggested that the 99% of the genome not coding for protein or RNA was "junk." Just as the earth is strewn with fossil remains of extinct species, so, he suggested, is our genome strewn with the remains of extinct genes (Ohno 1972). His characterization of 99% of the genome as "junk DNA" was generally accepted in the scientific community (e.g., Orgel and Crick 1980) and served to discourage the investigation of its possible functions (Makalowski 2003). This attitude characterizing DNA of unknown function as "junk" still lingered even as late as 1999. John W. Drake, of the Laboratory of Molecular Genetics, National Institute of Environmental Health Sciences, Research Triangle Park, NC, and formerly of the University of Illinois, who has done important research on mutation rates, wrote:

> "In most higher eukaryotes, much of the DNA either serves no function (other than to promote its own parasitism) or acts merely as a spacer." (Drake 1999)

Many researchers considered that investigating this "junk" would be as unrewarding as going through a garbage dump trying to discover some neglected valuables. The efforts, however, of those garbage hunters who did pick through the junk were rewarded with pearls of the discovery of remarkable unsuspected genetic and epigenetic functions, which are important components of my nonrandom evolutionary hypothesis (NREH) described in the next chapter.

Random point mutations, long considered the source of the genetic variety necessary to account for Common Descent, are incapable of playing that role and, as it turns out, are also incapable of accounting for much of the interesting evolution we observe today. Nonrandom genetic rearrangements, on the other hand, could account for this interesting evolution, as we shall see in the next chapter.

Chapter 2

The Nonrandom Evolutionary Hypothesis

E*PIGENETICS HAS BEEN CALLED ONE OF THE NEW BUZZWORDS IN* biology (Jablonka 2009). The term was first introduced by C. H. Waddington (1942) to describe the heritable changes in genetic functioning during embryonic development, but the term has come to mean more than that. An epigenetic trait has often been defined as a heritable trait resulting from a genetic change that is not a change in the DNA nucleotide sequence (Jablonka 2009). This definition reflects the use of the term to refer to the genetic changes during development, particularly those that are stimulated by inputs from neighboring cells during embryonic development. But this definition leaves out important phenomena such as what James Shapiro has called *natural genetic engineering* (1992, 1997, 2005). I think epigenetics should also include all genetic changes that are stimulated by the environment including those coming from outside the organism. All of these types of events comprise epigenetic events, and I shall so treat them here.

In my book *Not By Chance!* I introduced an hypothesis suggesting that much of the evolution we actually observe is the result of organisms' built-in capability to respond adaptively to environmental inputs. I called it the *nonrandom evolutionary hypothesis* (NREH). This kind of evolution relies on events that are epigenetic in the broad sense. It differs markedly from neo-Darwinian evolution, which postulates that a population changes when a rare random mutation takes over the population through natural selection. Neo-Darwinian evolution is said to be driven by random genetic changes, whereas the type of evolution I have suggested is driven by nonrandom epigenetic change triggered by environmental inputs.

I have suggested that an environmental change can cause the genome of an individual to be altered to effect an adaptive response to the change, and this altered form of the genome can be inherited. It is generally recognized that environmental inputs can stimulate epigenetic events, but it is not so generally recognized that a significant fraction of these are adaptive to the environment that did the stimulating. Animals and plants have the built-in ability to respond adaptively to environmental stimuli. This capability enables these plants and animals to adapt quickly to a changing environment.

The word *random* is often used when referring to the mutations that are the source of novelty in the neo-Darwinian evolutionary theory. That term means the occurrences of these mutations are independent of the effects they have on the phenotype. It is in this sense that mutations are said to be "random" in evolutionary theory.[1] This statistical independence, or colloquial "randomness," is what has served to distinguish the Darwinian theory of evolution from the hypotheses of the inheritance of acquired characteristics, the most famous of which is that of Jean Baptiste Lamarck. Darwin, himself, did not use the term *random* in describing heritable variation, but he has been understood

---

1 Evolutionists also use the word random to mean uniformly distributed. When they say that mutations are random, they often mean that all mutation rates are the same and if they are not the same they say it's not random. That usage is, however, not in keeping with the meaning of random in mathematics.

to have meant that — when he wasn't trying to co-opt Lamarck's ideas.[2]

I suggested this hypothesis because it could account for much of the rapid evolution that has been observed. The ability to respond requires that an organism be able to perceive a change in the environment and have a mechanism whereby that perception leads to the activation of a latent gene or other genetic resource, which in turn leads to a phenotypic change that will grant the organism an advantage in the new environment. Barry Wanner of Emory University has suggested something like that for bacteria — that genomic rearrangements could be part of a control system in bacteria that would produce heritable changes in response to environmental inputs (Wanner 1985). Christopher Cullis and his colleagues at Case Western Reserve University in Cleveland have reported that environmental changes induce genetic rearrangements in flax plants. The same genetic changes appear when the same environments are presented, indicating genetic changes are stimulated by environmental inputs and are not the products of chance (Cullis 2005, Chen et al. 2009). Although Cullis has noted his experiment was unable to show that the changes were adaptive to the given environment, he suggested that since the same changes appear in every individual under the same conditions, it is reasonable to suppose that the changes are indeed adaptive.

Antonio Prevosti and his colleagues at the University of Barcelona have found that the layout of the DNA in the fruit fly *Drosophila subobscura* varies with latitude. This variation does not appear to be arbitrary nor does it seem to be the result of chance, because the same variation with latitude is found both in North America and Europe (Prevosti et al. 1988). How can one understand this? The variation in the DNA can be understood as the possible result of an epigenetic response to the environment. The similarity of the variation of the DNA with latitude

---

2 In Darwin's sixth edition of *The Origin of Species*, he evidently found it necessary to invoke the Lamarckian principle of the inheritance of acquired characteristics. Although in the first five editions there is no mention of the inheritance of acquired characteristics, such mention appears five times in the sixth edition.

on the two continents indicates that in each case the layout of the DNA is related to the environmental conditions characteristic of the latitude and may be adaptive to them. Apparently, inputs from the local environment have been responsible for genetic rearrangements in the flies (Lee 2002).

In the last several decades there were already some biologists who felt that neo-Darwinian theory could not account for large-scale evolution (Ho and Saunders 1979, Shapiro 1992, 2009, Johnston and Gottlieb 1990). Noble (2013) claims the central assumptions of neo-Darwinism "have been disproved." I showed (Spetner 1997) that (1) speciation by the neo-Darwinian process is so highly improbable that it should be considered impossible, and (2) when random mutations were shown to produce some microevolution, they were not the kind of mutations that could lead to Common Descent even if they were to operate over an unlimited span of time.

Random point mutations, which neo-Darwinian evolution holds are the source of novelty in evolution, have not been shown to add any information to the genome. Usually, they have been seen to have lost information. I have stated (Spetner 1997) that no random point mutation has been observed that adds information to the genome, and the statement still holds. Some biologists are now beginning to realize that the genetic changes required for evolution have to be *nonrandom* (Shapiro 1992, 2009, Mattick 2009, Caporale 2003). Large nonrandom genetic changes are indeed known to occur, and these changes are under cellular control as noted in the previous chapter.

* * *

Nonrandom Epigenetic Events

The cellular mechanism that turns genes ON and OFF is now well understood according to the model of the *lac* operon in *E. coli* bacteria first described by Jacob and Monod (1961). A cell is capable of making thousands of different enzymes, and it makes them when, and only when, they are needed. When the bacterial cell senses the presence of sugar molecules that it needs, the cell typically turns ON the genes encoding the enzyme that transports the molecules into the cell and the enzyme that splits them apart into a form in which the cell's metabolic machinery can further break them down into energy packets (in the form of the molecule ATP) or use them to build structural elements. When the cell no longer needs the sugar molecule, or if it is no longer available, these genes are turned OFF. Thus, on the basis of external or internal inputs, the cell turns ON only the genes that are needed. This type of control is called *short-term*. The ON/OFF states of the genes are determined by the immediate needs of the cell. The bacterial cell, for example, is capable of feeding on a variety of sugars and it has built into it the chemical mechanism to switch from one to another as each becomes available. When more than one type of sugar is available, it will usually pick the one that is most efficient for it to use (Shapiro 2011).

I suggested in my previous book more than fifteen years ago that a cell, and even an organism, may have the built-in ability to also exercise *long-term control*, control that could span generations by adapting to long-term environmental changes. From research reports that have appeared since then, some of which I shall describe here, this suggestion is being vindicated. An environmental change can be a long-term challenge, and the organism can respond through a heritable change that will serve to adapt it and its progeny to the new environment. The organism can do this through an inherent, built-in capability to alter its genome to enable it to respond to the change. The cell may have other tricks it can do as well to accomplish the same purpose. This capability has some similarity to its ability to exercise its short-term control.

James Shapiro has suggested that cells have the capability of doing their own genetic engineering (Shapiro 1997, 1999, 2011). This capability is built into the cell, which enables organisms to alter their genome to adapt to a changing environment. Organisms thus have the capability to adapt quickly to a new environment.

We have no theory of how this capability became incorporated into the organism. The only evolutionary events that have been directly observed cannot account for it, nor can they account for Common Descent. No process is known through which the required information could have been built up. That lack of knowledge does not prove that the origin and development of life is necessarily a supernatural event, but it does show the emptiness of the present theories of the origin and development of life and of Common Descent.

Much has been learned in the life sciences in the last several decades about how an organism can alter its genome to enable it to adapt to new environmental conditions. Transposable genetic elements were discovered some seventy years ago by Barbara McClintock (McClintock 1941, 1950, 1955, 1956, 1983), but they were initially dismissed by mainstream geneticists as spurious phenomena. McClintock pursued her research despite it being considered a backwater area, and eventually the importance of her work was recognized by the Nobel Prize committee in awarding her the Prize in Medicine in 1983. The transposable genetic elements she discovered have been subsequently revealed to be members of a class of genetic rearrangements that do not occur spontaneously by chance but are under strict cellular control. In these rearrangements, sections of DNA can move from one place to another in the genome, or can be removed entirely. These controlled genetic changes can reveal latent genes that were in the genome but were previously unavailable to the organism. Hall (1999) has called them *cryptic genes*, but more about these genes later.

Environmental changes are known to elicit various kinds of stress in an organism. Furthermore, McClintock noticed in her early work in plants that some types of stress can trigger genetic rearrangements (McClintock 1984). Organisms seem to have the ability to relieve the stress by altering both their phenotype and genotype. *Stress* has been defined generally as an environmental condition threatening to upset the balance and stability of the organism. Stress in an organism includes many kinds of stimuli. In microorganisms, stress can be an excess or deprivation of necessary molecules, such as sugars or salts. It could also be excessively high or low temperature. In plants and animals, stress is usually a more complex form of environmental insult. Stress can elicit genetic rearrangements, which can in turn activate latent (or cryptic) genes. Many examples are known of genetic rearrangements activating latent genes (e.g., Shapiro 1992 & 2009, Hall 1999). Slack et al. (2006) have reported that stress can elicit an adaptive response in *E. coli* by selectively amplifying genes. Hersh et al. (2004) have reported that stress can induce adaptive genetic changes in *E. coli*. In subsequent sections I will discuss stress in both single-celled and higher organisms, and its heritable effects.

The nonrandom evolutionary hypothesis (NREH) I am suggesting is a paradigm very different from the random mutation of the NDT. I am suggesting an evolutionary process in which *individuals* evolve, as opposed to the neo-Darwinian process, which purportedly acts on *populations*. Stress can induce epigenetic changes in an organism, changes that can activate a set of latent genes in both its somatic cells and its gametes. An organism will often undergo stress if its environment changes significantly. The stress may then induce genetic rearrangements, which may turn ON some hitherto latent, or cryptic, genes. These latent genes may elicit a response in the organism that is adaptive to the new environment. The response is not a chance phenomenon. The adaptive response is a capability apparently built into the organism to match the environmental source of the stress. The number of potential responses would of course necessarily be limited, as would the number of environmental conditions that could elicit those responses. The genetic changes may occur in the entire population or only in some large fraction of it. If the change is adaptive, and if the changes occurred in the gametes, the population will soon consist of only those individuals that made the genetic change — an example of natural selection.[3]

---

[3] Note that I embrace natural selection but reject Common Descent, which cannot be accounted for by natural selection. Natural selection can be very effective when the adaptive genetic change occurs in many individuals rather than in only one.

Genetic rearrangements are mediated by repetitive sections of the DNA (Shapiro and von Sternberg 2005, Shapiro 2011). The most important DNA components for genome rearrangement are the transposons, which consist of integrated systems of proteins and nucleic acids. Transposons, which used to be known as "jumping genes," can change their location in the genome. They not only can jump around the genome themselves, but they can also control movement of other segments of the genome. They can replicate themselves and place the copy in another part of the genome (copy-and-paste), or they can just move without replicating themselves (cut-and-paste) (Shapiro 1999). These elements are the most abundant component of the human genome. They have been found to make up more than 42% of the human genome (Smit 1999). Compare this abundance to that of the protein-encoding DNA, which makes up less than 1% of the human genome.[4] Moreover, the type of repetitive DNA is specific to each category of organisms, differing from one category to another — differing even more than do the protein-coding portions of the DNA. In particular, in mammals, one important type of repetitive DNA, the Short Interspersed Nucleotide Elements (SINE), which are responsible for some of the DNA rearrangements, are known to be unique to each order of mammals (Shapiro 2002).

Nonrandom genetic changes triggered by an environmental input can account for the examples of evolution that have been actually observed (as opposed to inferred), whereas random mutation such as DNA copying errors cannot. Genetic rearrangements can produce the genetic differences that have been observed between closely related organisms. A cell has the ability to shift pieces of its genome around, sequestering genes that are not needed and revealing those that were previously hidden and that have promise to be adaptive in the current environment. Moreover, the genetic rearrangements that will reveal the adaptive genes are known to be triggered by inputs from the environment.

---

[4] I give the example of the human genome only because the most extensive genome study has been made on the human. I would expect that the same is true of all animal genomes, and perhaps of plant genomes as well.

An organism thus has the built-in ability to adapt to a new environment heritably by altering its DNA. These adaptations occur just when they are needed, because they are triggered by an input from the new environment. Since they are triggered by the environment, their occurrence in a population is not rare. They will occur in a large fraction of the population, leading to rapid evolutionary changes — possibly even in one generation! If such adaptive changes had to be achieved by random DNA copying errors (point mutations), they would require long expanses of time, if they could be achieved at all. It is not clear that there even exist long sequences of potential point mutations leading from the pre-adaptive to the adaptive form where each successive mutation grants to the phenotype greater adaptivity than what went before it. There may very well be no such potential sequences. It has never been shown that random point mutations could ever produce the required transitions even in unlimited time. Darwinists are always tacitly assuming such sequences exist. The burden of proof is on them to show it, and they have not and apparently cannot.

The epigenetic changes triggered by the environment are nonrandom, and are unlike random point mutations. When they appear in a population, they appear in many individuals in the same generation. They can appear in many members of the population of organisms experiencing the changing environment. Although a rare point mutation is very likely to be lost from the population even though it may have substantial positive selective value (as noted in Chapter 1), the presence of many copies of an adaptive mutation in a population will be almost certain to increase their numbers through natural selection and take over the population. If these genetic changes were to be adaptive, they would very quickly take over the population through natural selection and the new adaptive genetic change would quickly become a characteristic of the population.

* * *

Conserved DNA

Strings of DNA are called "conserved" if the same sequences appear in different species of organisms. The term "conserved" stems from the evolutionary paradigm under which Common Descent is assumed. If the same DNA sequence appears in organisms of different species, genera, families, or phyla, it is assumed the sequence was "conserved" in an evolutionary process by natural selection while other parts of the genome were changed through random mutations. Consider two species of organisms that, under the evolutionary assumption, have descended from a common ancestor in the distant past, and which have some identical matching strings of DNA bases. The Darwinist concludes that these sequences must have been serving a useful purpose throughout the generations since the time of the alleged common ancestor, and therefore they were kept intact, or "conserved," by natural selection as other parts of the DNA suffered random mutations. But under any assumption, evolutionary or not, if the sequences appear very similar in two very different organisms, they must be serving some important purpose in both.

If these DNA strings are genes, their purpose is evident — they code for protein or RNA. Interestingly, though, the most "conserved" strings are *not* genes. A large number of nongenic sequences have been found that are even more strongly "conserved" than the genes. They must therefore have a very important function other than presently coding for protein or RNA.

These conserved sequences have been given the name *conserved nongenic* sequence (CNG) (Dermitzakis et al. 2005). A sequence is considered to be a CNG if it matches at least 70% in two species, say between mouse and man, over a length of at least 100 base pairs. This quality of match is somewhat better than what is found on average between the human and mouse genome. Between 1% and 2% of the genome qualifies to be CNG, wheras the genes occupy about 1%.

There are also sequences in the genome that have an interspecies match even better than that of the CNG's and are called *ultraconserved*. These are defined as sequences of at least 200 base pairs that are 100% identical from mouse to man (Bejerano et al. 2004). They must be serving some important function. They also are not genes; so what could be their function?

Nadav Ahituv and his colleagues at the Lawrence Berkeley National Laboratory in Berkeley, California, did an experiment designed to discover the function of sequences that are ultraconserved between mouse and man. They deleted four such sequences from the genome of mouse embryonic stem cells to see what effect their elimination would have on either the embryo development or the mouse function (Ahituv et al. 2007). To their surprise they found the mice whose ultraconserved sequences were eliminated were perfectly normal; the knockout of the ultraconserved sequences had no effect on the mice — neither in their function nor in their development! These sequences must have a function, however, under the evolutionary assumption or under any other reasonable assumption. What can that function be? That function, no matter how important, must be unrelated to the development or the day-to-day needs of the mouse. It is reasonable to suggest their function may be related to the ability of the organism to evolve its response to environmental changes.

There is experimental evidence that there are DNA sequences that have no effect on the functioning of the organism under ordinary circumstances but which play a role under extraordinary conditions. An experiment was performed on yeast in which each of its 6,000 genes was deleted, one by one (Hillenmeyer et al. 2008). Of these 6,000 genes, 34% were found to be necessary for the proper functioning of the cells under normal conditions because their deletion was either lethal or led to growth defects. The remaining 66% of the deletions *showed no effect* under normal conditions! Almost all of these (63% of the total), however, *showed growth defects under various environmental changes*! The remaining 3% showed no effects in this experiment. It is possible, though, they would have shown some growth defects under some other environmental conditions that were not tested. Thus, some two-thirds of the genes studied are likely to be the part of the genome containing the yeast cell's built-in ability to adapt to environmental changes. There is thus good evidence that a significant fraction of the genome is dedicated to adapting the organism to changing environmental conditions.

* * *

Nonrandom Evolution of Cells

Nonrandom evolution, as in the NREH, has been reported to occur extensively in single-celled organisms, and in particular in bacteria. Some evolution in bacteria is known to occur rapidly — much too rapidly to be accounted for by random errors in copying DNA. But it *can* be accounted for by nonrandom, adaptive, genetic rearrangements triggered by environmental inputs.

Cryptic Genes

Cryptic genes, to which I referred above, are genes that are normally silent during the life cycle of an individual but can be activated by an epigenetic change (Hall et al. 1983). A cryptic gene has all the characteristics of an ordinary gene with all its control apparatus, except that it is prevented from being active by a DNA segment called a silencer, which does not allow it to be turned ON (Schnetz 1995). This silencer is to be distinguished from the repressor of an ordinary gene, which keeps the gene OFF until an inducer protein turns it ON.[5] In fact a cryptic gene has its own inducer and repressor and is fit to operate fully like an ordinary gene except that the silencer won't let it.

If its silencer were to be turned OFF, the cryptic gene would function like an ordinary gene. A silencer is sometimes turned OFF by an insertion of a segment of DNA (Reynolds et al. 1986 as cited by Parker et al. 1988, Schnetz 1988, Parker and Hall 1990) or by the deletion of

---

[5] For a description of the control of a gene by the repressor and the inducer, see Appendix H of *Not By Chance!* or any good high school biology text.

a segment of DNA (Parker et al. 1988, Schnetz 1995, Schnetz and Rak 1992). The silencer does not turn OFF at random — it is under cellular control.

Cryptic genes have been found in most bacteria and may be found in higher organisms, as well. For example, at least 90% of the bacteria *E. coli* have cryptic genes for β-glucoside sugars (Hall and Betts 1987 as cited by Hall 1999). Cryptic genes have been reported that code for the enzyme acetohydroxyacid synthetase, which catalyzes the first step in the synthesis of the amino acids valine, leucine, and soleucine (Mukerji and Mahadevan 1997). The heat-shock protein Hsp90 has been found to cause the expression of a cryptic gene in Drosophila when stimulated by an elevated environmental temperature (Rutherford and Lindquist 1998).

Cryptic genes are also known that code for antibiotic resistance in bacteria (Hall 2004). These cryptic genes are generally found on plasmids,[6] and when no antibiotics are present the cryptic genes are silent, meaning that the resistance is not expressed in the microorganism. In the presence of antibiotics an epigenetic change can occur, turning OFF the silencer and permitting the cryptic resistance genes to be expressed. This means that the antibiotics we administer against bacterial infection are inducing their own resistance. We have to discover what in the antibiotic is doing that and try to design the antibiotic around it.

The environment elicits the insertions or deletions that disable the silencer to allow the cryptic genes to be expressed. Note that these epigenetic changes are regulated, occurring when the organism is being stressed by its environment (Hall 1999). Hall described the importance of insertion elements in adaptively modifying the genome:

> "Not only do *IS* elements make a direct contribution to fitness by activating cryptic operons, they do so in a regulated manner, transposing at

---

[6] A plasmid is a cellular element containing DNA, most often found in bacteria, that replicates independently of the chromosomes. Many plasmids have from 50,000 to 200,000 base pairs of DNA.

a higher rate in starving cells than in growing cells. In at least one case, *IS* elements activate an operon during starvation only if the substrate for that operon is present in the environment."

The activation of a cryptic gene is an example of a genetic rearrangement responding to environmental inputs to make the organism more adaptive to the new environment.

The "Nylon Bug"

Nylon was invented in a DuPont research laboratory in Wilmington, Delaware, in 1935 and began to be produced commercially in 1939 to be used to make ladies' stockings. Nylon has since been used to make a variety of fabrics, and for high-strength fiber for ropes, nets, and parachutes. In 1975 a bacterial strain (*Flavobacterium*) was reported discovered in a pond of wastewater of a Japanese nylon factory. The bacteria were living off the nylon waste, which was, of course, new to the earth's biosphere (Kinoshita et al. 1975). Japan started producing nylon only in 1951, so these bacteria had no more than about 20 years to obtain their new capability. The bacteria were obtaining all their carbon, nitrogen, and energy from the nylon waste products (the energy needed for the bacteria comes from the potential energy in the chemical bonds of the nylon waste molecules). These bacteria were found to have three new enzymes that together were able to metabolize the nylon waste. Before the era of nylon, the *Flavobacteria* had no need for these enzymes. The enzymes were tested against about 100 natural molecules similar to the nylon products and they were unable to catalyze any metabolic reactions with them (Kinoshita et al. 1977, 1981). The enzymes evidently appeared in response to the stress of a dearth of their normal nutrient in the presence of an abundance of the nylon waste. The bacteria somehow developed (evolved?) enzymes necessary for them to metabolize this waste product.

One of the components of the waste from a nylon factory, called *6-aminohexanoic acid cyclic dimer (Acd)*, is a combination of two molecular chains of six carbon atoms and a nitrogen atom, attached to

CYCLIC DIMER

```
      H H H H H
      | | | | |
O=C - C-C-C-C-C - N-H
      |           
      H H H H H     Enzyme 1
      |           breaks this
                  bond
      H H H H H
      | | | | |
H-N - C-C-C-C-C = O
      | | | | |
      H H H H H
```

⇩

LINEAR DIMER

```
               H H H H H H
               | | | | | |
         O=C - C-C-C-C-C - N
               | | | | | |
Enzyme 2       H H H H H H
breaks this
bond
               H H H H H
               | | | | |
         H-N - C-C-C-C-C-C-O-H
               | | | | | ‖
               H H H H H O
```

⇩

2 MONOMERS

```
  H H H H H H
  | | | | | |
N-C-C-C-C-C-C-O-H
  | | | | | ‖
  H H H H H O

  H H H H H H
  | | | | | |
N-C-C-C-C-C-C-O-H
  | | | | | ‖
  H H H H H O
```

Fig. 2.1 The nylon waste water contained the cyclic dimer molecule shown at the top. Two monomers are joined to each other at both ends by chemical bonds, forming a closed loop or cycle. Enzyme 1 breaks one of these bonds, forming a linear dimer. Enzyme 2 then breaks the second bond severing the two monomers from each other.

each other at both ends forming a cyclic configuration as shown in Fig. 2.1. This molecular arrangement is called a *cyclic dimer*. It's called a *dimer* because it has two linear monomer components, and *cyclic* because the two linear components are attached to each other at both ends forming a loop or cycle. The bacterium metabolizes the cyclical waste molecule in stages as is customary with most biochemical reactions. The first newly developed enzyme (designated as E1) hydrolyzes the dimer, catalyzing the breaking of one of the two bonds attaching the two monomers by adding a water molecule, thus opening the closed cycle, as shown in the figure, to form a linear dimer called *6-aminohexanoic acid linear dimer (Ald)*. Then the second enzyme (E2) hydrolyzes the linear dimer by catalyzing the breaking of the other bond joining the two components by adding another water molecule to form two monomers. The bacterium already had the necessary enzymes to metabolize the monomers. The genes encoding these two enzymes are on one of three plasmids in the bacterial cell and their DNA sequence has been determined (Kinoshita et al. 1977, 1981). The genes encoding E1 and E2 have been denoted as *nylA* and *nylB* respectively. A third enzyme (E3) was also discovered (Kakudo 1993) that hydrolyzes other components of the nylon waste, and its gene, denoted by *nylC*, was also found on the same plasmid in *Flavobacterium* as were those of E1 and E2.

The appearance of these genes in *Flavobacterium* is evidently not the result of the random DNA copying errors of Darwinian evolution. Note that both genes, *nylA* and *nylB*, had to evolve for the bacterium to be able to metabolize the *Acd*. Just one of them would have been of no use. The appearance of the two genes seems to be the result of a built-in capability of the bacterium, enabling it to respond to an environmental input as the NREH postulates.

It turns out that *Flavobacterium* is not the only bacterial strain in which a genetic change occurs to enable it to metabolize nylon waste. Another bacterium was discovered in which the same genetic change has occurred. Seiji Negoro and his colleagues at Osaka University, Japan, isolated a strain of the bacterium *Pseudomonas* in which a *nylA* gene and a *nylB* gene also appeared that coded for the two enzymes, E1 and E2 respectively, that allowed the bacterium to metabolize the cyclic dimer from nylon waste (Kanagawa et al. 1989). The E1 enzyme of *Pseudomonas* is 99% identical with that of *Flavobacteria* (Tsuchiya et al. 1989), while the E2 enzymes are less similar with only a 35% correspondence (Kanagawa 1993). The two enzymes in *Pseudomonas* are located on two different plasmids, whereas in *Flavobacteria* they are on the same plasmid. (The E3 enzyme has so far not been detected in *Pseudomonas*.) The 99% correspondence of the E1 enzymes shows their appearance in *Pseudomonas* was unlikely to be just the result of random DNA copying errors.

Was the appearance of the nylon genes a lucky evolutionary event for the bacteria, or is it an event that is likely to happen whenever the nylon-waste molecules are present for the bacteria to exploit? Does the appearance of the nylon genes in both *Flavobacteria* and *Pseudomonas* indicate these bacteria have an inherent capability to create or reveal these genes? Negoro's group at Osaka University experimented with a strain of *Pseudomonas* that could not metabolize the nylon waste and cultured it with the nylon waste as its only source of carbon and nitrogen (Prijambada et al. 1995). These experimenters took a strain of *Pseudomonas* from New Zealand, which is more than 11,000 miles from the Japanese nylon factory where it was very unlikely to have been contaminated from the Japanese *Flavobacteria*. They cultured the *Pseudomonas* in the presence of *Acd* as the only source of carbon and nitrogen. *Within only a few months*, a strain appeared in the culture having the *nylA* and *nylB* genes encoding the E1 and E2 enzymes that would metabolize the *Acd* and permit the bacteria to live off these molecules.

How did these bacteria acquire the genes necessary for them to live in their new environment? How did they do it so quickly? They could not have done it through random errors in DNA copying, which even if it could work would require enormous spans of time. They must have done it through epigenetic changes that were triggered by the presence of *Acd* and a dearth of other nutrients. The authors of the study wrote:

> Though a molecular basis for the emergence of nylon oligomer metabolism in [the bacteria] is still unknown, it is probable that the basic mechanisms acting during environmental stress are involved in this adaptation. (Prijambada et al. 1995)

The enzymes that can metabolize the nylon waste appeared in response to the stress of starvation in the presence of the waste and were likely triggered by an epigenetic event that was itself triggered by that stress.

* * *

Nonrandom Evolution in Higher Organisms

Stressful conditions in higher organisms usually stem from conditions more complex than those affecting microorganisms. The effects of such complex stressful conditions are usually mediated through the brain. Examples of environmental stress include the appearances of predators, overcrowding in a habitat, and severe climatic change. In fish, stress is known to lead to changes in both behavior and physiology (Iwama 1998). These changes usually serve to reduce the stress on the fish. Complex environmental inputs may trigger the epigenetic changes according to the NREH. Iwama divides the stress response in fish into three stages. The primary response is the release of special stress hormones (e.g., catecholamines and ACTH) into the blood. The hormones are generally referred to as *first messengers*. These hormones then travel through the blood stream to all cells in the body, activating sensors on the surface of the cells. These sensors trigger the secondary response, effected by molecules within the cell, generally referred to as *second messengers*. The second messengers then trigger appropriate enzyme activity (Cohen 1988), which can affect both physiology and behavior, and which are apparently designed to reduce the effects of the stress. These second messengers can turn genes ON or OFF. The genetic changes may be manifest in the individual in some of its life processes such as growth and reproduction, or it may be manifest in the population as a whole in its abundance and diversity.

* * *

The NREH can result in a newly evolved species within the same living space as the original species, called *sympatric speciation*; this may be the way many instances of sympatric speciation occur. Many of the examples of rapid evolution have turned out to be examples of sympatric speciation, with the appearance of an isolating mechanism that prevents the new species from being diluted by the extant species. Although the concept of sympatric speciation has gone in and out of favor among students of evolution over the past century and a half, evolution *as observed in action* does show sympatric speciation.

For a long time, sympatric speciation was thought to be an important method of speciation until Ernst Mayr (1963) gave cogent theoretical arguments against it. He argued, on the basis of neo-Darwinian theory, that sympatric speciation could occur only with the simultaneous appearance in the nascent species of a preference for a new niche and cross-sterility with the extant species, or at least a preference for mating with its own kind, a pair of occurrences unlikely to appear randomly and independently. Under the neo-Darwinian paradigm, Mayr was of course perfectly right. In addition, he held that, at that time, the data supporting sympatric speciation were not irrefutable, which allowed him to ignore them.

By the end of the twentieth century, however, enough irrefutable examples of sympatric speciation had been accumulated that Mayr was compelled to acknowledge its existence and importance (Mayr 2004). In the latter half of the twentieth century, many more examples of rapid sympatric speciation were discovered. See, for example, Schoener (1970), Barton et al. (1988), Feder et al. (1988), Dieckmann and Doebeli (1999), Schilthuizen (2001), Via (2001), Savolainen et al. (2006), and Barluenga et al. (2006). Moreover, Thoday and Gibson (1962) claimed to show the evolution of sympatric speciation in the laboratory. With the abundance of examples of sympatric speciation, Darwinists felt compelled to devise a theoretical basis for it, which of course they strived to contain within the Darwinian paradigm. The late John Maynard Smith (1966) proposed a theoretical model of sympatry. In the following decades, more models of sympatry were suggested. Some of the models postulated the immigration into an existing population of a few individuals already sexually isolated from the extant population (Doebeli 1996). This scenario could of course happen, but it is not likely that it could account for most of the observed examples of sympatric speciation. Others postulated the appearance of a few reproductively isolated mutants, but such mutants, if they occur at all, must be very rare. Their appearances are nearly impossible events under the neo-Darwinian paradigm. None of the models that were proposed to account for sympatric speciation based on neo-Darwinian theory can account for the data. The many examples of sympatric speciation do not fit with neo-Darwinian evolution. They can be explained by environmentally-induced epigenetic events of the NREH, but not by random mutations and natural selection.

* * *

For centuries, naturalists have observed that populations of animals and plants often change with a changing environment. Proposed explanations for this phenomenon have generally involved the inheritance of acquired characteristics (Waddington 1942), of which Lamarck's suggestion is perhaps the most famous. Note here that although Lamarck's theory is often ridiculed in biology textbooks, he was no fool. From the large volumes he wrote on zoology, one can see he was an expert on the natural history of animals. He was undoubtedly familiar with many examples of animals changing to adapt to new environmental conditions. He tried to account for the morphological and behavioral changes he observed to occur as a result of the animal's need for the change. Explanations of speciation based on the inheritance of acquired characteristics were ultimately rejected because there was no adequate evidence for it and no known mechanism by which it could occur (Waddington 1942).

J. Mark Baldwin (1896) suggested an explanation of a special set of phenomena wherein adaptive behavior arising from an environmental change can become hereditary. His speculation was that an environmental change stimulates animals to adopt an adaptive behavior that helps them adjust. The next generation learns the behavior by observing their parents. Some are better learners than others and the better learners fare better in the new environment than the poorer ones because the learned behavior is adaptive. He reasoned that the ability to learn would therefore have selective value and be enhanced in the population through natural selection. Although the behavior itself is not heritable, he thought that the ability to learn might be heritable and that natural selection would therefore have something to work on. However, this "Baldwin effect," as it has been called, cannot account for the evolution of anything other than behavior. Even then, of course, it suffers from the general problem with neo-Darwinian evolution in that it does not explain how the new behavioral information built up, except to say that perhaps random mutations did it.

Another suggestion to explain what looked like the heritability of acquired characteristics was offered by Conrad H. Waddington (1942) of the University of Edinburgh, who was renowned for his contributions to embryology and genetics. He was intrigued by the many examples of what seem to be the inheritance of acquired characteristics. In particular he addressed the sternal calluses on the ostrich as described by Duerden (1920). These calluses are on tissues that rub or press against an external surface. Although the calluses can be formed from the friction of rubbing, they actually develop in the embryo before the tissue encounters an external stimulus. The Darwinists' problem with the two ways these calluses can develop is that it looks too much like the inheritance of acquired characteristics and they knew of no mechanism that could account for it.

Waddington's suggestion was essentially a scenario that, in the early ostriches, the calluses were formed only as a response to the external stimulus, but "during the course of evolution the environmental stimulus has been superseded by an internal genetic factor." To fit his suggestion into the Darwinian paradigm, one must show how a genetic mechanism can arise from random mutations and natural selection. Even leaving aside the issue of the existence of a sequence of increasingly adaptive random mutations, how can natural selection have an effect? From the beginning, the environment stimulates the formation of the calluses, so what advantage is there to the appearance of a genetic source to bring on the same calluses? Natural selection cannot distinguish between calluses brought on by environmental stimuli and those that are congenital.

The NREH, however, can account for the calluses both appearing as a result of an environmental stimulus and developing in the embryo. The formation of the calluses in the ostrich is under genetic control (Waddington 1942). The pressure and friction against the skin is a kind of stress, and stress is known to be able to stimulate the release of hormones. In many cases these hormones yield an adaptive response to the stimulus that can act on the genetic control of the callus formation. These hormones reach all cells; they reach the gametes as well as the somatic cells and affect the same region in both. Since they reach the gametes, their effect is heritable. Although this description is somewhat speculative, it is based on known phenomena; it is far less speculative than the assertion of the neo-Darwinian theory that random mutations can produce adaptive genetic changes when they are needed.

* * *

The ability of an organism to alter its phenotype in response to an input from the environment is a well-known phenomenon, referred to as *phenotypic plasticity*. The phenomenon is considered *plastic* because it is believed that when the environmental input is removed, the phenotype returns to its previous state, which means it is not heritable. It turns out, however, that these phenotypic changes can nevertheless be heritable, indicating they also have a genetic basis.

Bradshaw (1965) considered plasticity in plants, but only the plasticity under genetic control. He did not consider the possibility that there may be epigenetic changes in the genotype as well as changes in the phenotype. There had been a strong tendency to dismiss as an irrelevant oddity the apparent influence of the environment on phenotypic changes, and certainly on the heritability of such changes. Mary Jane West-Eberhard (1989) described some manifestations of this attitude. She wrote:

> In the 1960s Wigglesworth (1961, p. 107) described some geneticists as being "apologetic" about environmentally cued polymorphisms, which

they considered examples of unfortunate defects in the delicate genetic apparatus: "As R. A. Fisher once said to me, it is not surprising that such elaborate machinery should sometimes go wrong." And Bradshaw (1965, p. 148) noted that botanists were carefully avoiding any mention of plasticity; environmental effects in experiments were considered "only an embarrassment."

West-Eberhard (1989) argued that the environmental influences were strictly plastic and did not involve genetic change. Later, however, in the face of much evidence, she acknowledged that somehow genetic change could indeed follow the appearance of environmentally induced phenotypic change (West-Eberhard 2005). She did not explain, however, how that could happen. Today environmental influences on the morphology and behavior of animals and plants are known to occur and be heritable.

Many reports of examples of rapid adaptive changes have appeared in the last several decades. These examples have been hailed as "evolution in action," and the hailers always mean neo-Darwinian evolution — random mutations and natural selection. But no one has addressed the problem of how random mutations could have produced these rapid changes, and it seems that indeed they cannot.

When I proposed my nonrandom evolutionary hypothesis (NREH) about fifteen years ago in my book *Not By Chance!*, the suggestion — that environmental inputs could affect the genome, leading to the possibility that an organism can change its physiology and behavior in response to an external input — was somewhat sketchy. A distinguishing feature of a good theory is that evidence for it grows after it has been formulated. Since I suggested the NREH, the following discoveries have been made that provide a solid mechanism for it, showing how organisms can show an adaptive heritable response to environmental inputs:

- Environmental inputs can cause stress in an organism.
- Stress can cause genetic rearrangements, which can be adaptive.

- Stress can induce the production of hormones, which can reach every cell in the body.
- Hormones, as first messengers, can activate sensors on cells.
- This activation triggers the second messengers inside the cell, which can bring the message to the DNA.
- The second messengers may well be the proximate trigger for the genetic rearrangements that ultimately are caused by stress.
- Genetic rearrangements can produce changes in the chemical functioning of the cell, which can in turn affect the physiology and behavior of the organism. These changes can be adaptive to the new environment.

These are now known phenomena through which the NREH can function.

* * *

Pupfish Evolution

Here is a good example of the NREH in action. The desert of Death Valley is the driest and hottest place in North America. In spite of the heat and dryness it is home to a unique type of fish, albeit a strange one — a unique species of the pupfish. These pupfish have shown their ability to evolve rapidly in response to environmental changes. Death Valley is home to nine species and subspecies, each in its own habitat and all isolated from each other. Their evolution has been compared to the evolution of the finches of the Galápagos Islands. Evolution in the pupfish is most noted in their body shape and behavior, whereas in the finches the evolution most noted is in the beak shape (Lema 2008).

In all of Death Valley, the most extreme environment in which pupfish are found is in Devil's Hole. Its high temperature and meager food supply are a challenge to the resident pupfish. Its pupfish are unique to Devil's Hole, and they are considered to be one of the world's rarest vertebrates. They are considered to be an endangered species because they live only in Devil's Hole and would disappear if their habitat were to be seriously altered. In a program to save endangered species, some pupfish were taken from Devil's Hole and moved to three special pupfish refuges. Although the refuges were constructed to match the conditions in Devil's Hole, the match could evidently not be perfect. After only five years, the pupfish morphologies in the refuges were observed to have diverged from that in Devil's Hole.

The divergence was thought to be the result of common built-in phenotypic plastic responses to similar environmental conditions. To test the possibility of phenotypic plasticity, Sean Lema and his colleagues took newly hatched pupfish from elsewhere (the Amargosa River, an intermittent waterway in Death Valley) and reared them under conditions made to mimic conditions in Devil's Hole. They found that even slight differences in rearing temperature affected the development and resulting morphology of the pupfish. They found that environmentally altered production of a thyroid hormone was a cause of the change in morphology. They also found that the hormone arginine vasotocin (AVT) plays an important role in altering the behavior of the pupfish when the environment is changed. The cells that produce AVT are located in the brain and are known to respond to stress.

From the studies on the pupfish that were taken from the Amargosa River, it was found that environmental changes induce stress that affect the production of hormones, which in turn affect the morphology of the pupfish. Moreover, from the studies of the pupfish that were moved from Devil's Hole to the refuges, it was found that the DNA of those in the refuges differed from those in Devil's Hole. These studies of the Death Valley pupfish offer substantive support for the NREH.

* * *

In addition, an increasing number of examples of population changes are coming to light that occur with a speed that cannot be accounted for by DNA copying errors or by the neo-Darwinian mechanism of random mutations and natural selection. Yet these examples are well accounted for by the mechanism of the NREH. Moreover, neo-Darwinian theory has always been quite vague about how the long strings of mutations they require can happen at just the right times, in which each mutation in the sequence has selective value over the previous one. In the next chapter I will provide several examples of rapid evolution observed today that the neo-Darwinian paradigm is at a loss to account for, but which fit very well with the NREH.

The NREH explains the inheritance of adaptive responses to stressful environmental inputs as the working of a built-in capability of the organism to adapt to long-term changes in the environment. It does not address the issue of the origin of this ability. How did this ability develop? Currently, we don't know, so this issue must be left for further study. But to give the Darwinist's glib answer that it arose by random mutation and natural selection without explaining in detail how that might have happened is not an answer.

Chapter 3

Rapid Evolution Is Happening Today!

In this chapter, I will examine several representative examples of evolution happening and observed today and, as we shall see, none of these examples offers any support for Common Descent. Some of these examples can be accounted for by the NREH. Only from direct evidence, such as these examples, can one hope to understand how evolution really works and to what extent, if any, the concept can be extended beyond our observations to make inferences about Common Descent. Evolution in the sense of adaptive population change does occur, but it is faulty science to infer from examples of those changes that Common Descent has occurred. In the examples I shall cite here, it will be clear why evolution of this sort does not support Common Descent.

* * *

Daisies, like many weeds, have wind-dispersible seeds. Each tiny seed is surrounded by a much bigger ball of very light fluff, which provides it with a large surface area and adds very little weight. To the seed are attached many fine hairs which, when the seed is released, open up into a large light feathery ball. The seed is thus equipped with something like a parachute to keep it afloat in the air. The seeds can be transported long distances even in a mild breeze. This seed-dispersal mechanism provides the daisy with the ability to send its DNA over long distances, giving it a reproductive benefit in spreading itself over a large area.

But sometimes this benefit can become a liability. On a small ocean island, the daisy's seed-dispersal ability would cause most of its seeds to be lost at sea. To counter this liability, the daisy evolves to stop dispersing its seeds when it becomes a disadvantage. Martin Cody and Jacob Overton, then at UCLA, studied more than 200 islands of various sizes harboring daisy-like plants with seed-dispersal ability (Cody and Overton 1996). They discovered that when these kinds of plants are transported from the mainland to small islands, they lose their seed-dispersal ability in just a few years. These changes in the plants are genetically-based and are heritable.

The daisy's ability to make the genetic changes enabling it to dispense with its seed-dispersal ability is an example of rapid evolution, which lends no support for Common Descent. This seems, rather, to be an example of the NREH, wherein an environmental input triggers an epigenetic change that is adaptive to the new environment. This ability cannot be the result of randomness, but may be a controlled response to an environmental input. Alternatively, as in many of these cases, one could suggest that the genetic configuration for the loss of seed-dispersal ability had always been in the population and when conditions are favorable for it, it comes to the fore. Neither of these two possible mechanisms can add any information to the daisy population's collective genome. In the first case, the capability is built into the genome and its functioning requires only a trigger from the environment to bring it out. In the second case, there are two distinct genomes in the population, but both were there from the beginning. The daisy example thus offers no support for Common Descent, even though the example qualifies as evolution under its definition as mere population change.

Curiously, Professor Jarod M. Diamond, of the University of California Medical School, thinks the work of Cody and Overton provides what he considers a rebuttal to Creationists who claim evolution has never been observed (Diamond 1996). Cody and Overton do indeed provide a good example of rapid evolution and a seemingly good indication of the power of natural selection as they claim. But it is support only for the type of evolution that comes under the definition of population change. It offers no support for evolution in the sense of Common Descent. Incidentally, Cody and Overton themselves make no claim of support for Common Descent. Diamond notes "that the large-scale evolutionary changes of most interest to us simply cannot be followed, because they take place over times far longer than a human lifespan." It therefore follows that we must gain our information about evolution from the examples that are rapid enough for us to observe. But the many examples of rapid evolution are evidently driven by mechanisms that cannot produce Common Descent *regardless of how long they act*. None of these examples can support Common Descent.

One must try to avoid confusing rapid evolution (often called *microevolution*), in which adaptive genetic rearrangements are made, with what Darwinists believe is the grand sweep of evolution known as Common Descent. I am reminded of a little story told to me by the late Dr. Judd Blass.

Mr. Jones makes a phone call and a four-year-old boy answers.

"May I please speak with your father?"

"He can't come to the phone now. May I take a message?"

"This is Mr. Jones calling."

"Okay."

"Do you know how to write?"

(Hesitantly) "… Yeah."

"Write 'Jones.'"

"Uh ... How do you spell it?"

"J-O-N-E-S"

... (A short period of silence)

"Are you writing it?"

"Yeah ..." (pause) "... Uh ...how do you make a 'J'?"

The four-year-old can write, but he needs help in making a "J." He is a writer, and Shakespeare is also a writer. There is one kind of writer whose struggle is with transcribing his thoughts into well-written prose or poetry. There is the other kind of writer whose struggle is with how to make a "J." Just as one does not confuse these two kinds of "writers" with each other, so one must not confuse Common Descent with mere change in populations, even though both are called *evolution*.

* * *

The apple maggot fly *Rhagoletis pomonella* originally bred and fed on hawthorn, but in the nineteenth century the fly left the hawthorn and began to infect apple trees. It has now spread to breed and feed on cherries, roses, and pears. When these flies took on a new host, they underwent several behavioral modifications:

- Their preference changed from the old host to the new host.
- They adopted new mating preferences of both the males for the females and the females for the males, which helped to isolate them reproductively from the old population.
- Their mating procedures changed to isolate further the new changed population from the old and keep them from interbreeding.
- They adjusted their maturation time to match the ripening of the new host fruit.

The second and third of these modifications yielded the reproductive isolation necessary for sympatric speciation, as described in Chapter 2. All the above modifications have been shown to be genetically-based and are heritable (Barton et al. 1988, Feder et al. 1988, McPheron et al.

1988, Smith 1988). Because these behaviors are genetically-based, they require changes in the genome for the manifestation of their modifications. All these features had to change quickly and simultaneously, something that a sequence of random errors in DNA replication, each followed by natural selection, cannot accomplish. But a built-in response capability, as postulated by the NREH, can do it. Filchak et al. (2000) have suggested that the warmer temperature inside the apple may induce an earlier maturity. Perhaps that is so, but McPheron et al. (1988) have reported that the hawthorn and apple flies differ genetically, which seems to favor the NREH explanation. Moreover, the reproductive isolation seems to rule out the possibility that both types of flies were in the population to begin with. In any case, this example of evolution-in-action shows no increase of information in the genome, and therefore offers no indication of how information might be built up in Common Descent. The example lends no support for Common Descent because any of the explanations of how this evolution might have occurred do not involve processes that could build up information in the DNA no matter over how long a time it might operate.

* * *

Almost anyone with a small fish tank in his house is familiar with guppies. These small fish have been reported to exhibit rapid evolution (Reznick et al. 1990, Reznick and Bryga 1987 & 1996, Carroll et al. 2007, Gordon et al. 2009). These small fish are preyed upon by larger fish that share the same living space. Guppies, it turns out, adapt their morphology and behavior to the type of predator in their area. Two types of such adaptations have been studied. Cichlid fish prey on large mature guppies and killifish prey on small immature ones. In the presence of cichlids, guppies adapt by maturing early and having many small offspring, which tend to evade the cichlids. In the presence of killifish, guppies tend to mature late and have fewer but larger offspring, which tend to evade the killifish.

To study how these adaptations evolve, David Reznick and his colleagues at the University of California have studied guppies in the Aripo River in Trinidad. This river has guppies together with cichlids, which prey on the large mature guppies. A tributary of the Aripo has killifish but had no cichlids and, until Reznick and his team came, it had no guppies. Reznick took 200 guppies from the Aripo and transferred them to the tributary. Changes soon appeared in the newly introduced guppy population. The population soon changed to what would normally be found in the presence of the killifish, and Reznick found the changes to be heritable.

The full change in the guppy population was observed as soon as the first samples were drawn, which was after only *two years* — much too short for random mutation and natural selection to have an effect. Reznick interpreted these changes as the result of natural selection acting on variation already in the population. This means no new information was generated in this example of evolution — the "novelty" had already been in the population. If so, this example of evolution cannot support Common Descent, which requires the generation of new information through random mutations. It is also possible the novelty was generated by genomic changes induced by the environment, as in the NREH. In either case, there is no support here for Common Descent.

* * *

Lizards have been observed to have evolved to adapt to a new environment in only ten to fourteen years (Losos 2001, Losos et al. 1997, Case 1997). Lizards were introduced on several islands in the Bahamas (Schoener and Schoener 1983). After about ten years they were observed to have diverged to become adapted to various niches, thus showing rapid speciation. Here, too, the changes were much too rapid to be accounted for by DNA copying errors and natural selection. What is interesting about these lizards is that the same diversity of lizard species is present on the four islands of the Greater Antilles: Puerto Rico, Cuba, Hispaniola, and Jamaica. That, however, is not the most interesting part. If the evolution of these species stemmed from random point mutations and natural selection, one would expect these species would have had to have that diversity within their population before they were distributed among the islands, and this same diversity would have then spread to the four islands. DNA analysis, however, shows that the species diverged in a parallel fashion independently on each island (Losos 2001). Such parallel evolution is highly unlikely to stem from random mutations. I have shown previously that convergent and parallel evolution through random mutations is so highly improbable that in practical terms it must be considered impossible (Spetner 1997). (Further discussion of convergent and parallel evolution is in Chapter 4.) This kind of evolution would, however, be expected on the basis of genetic changes triggered by inputs from the environment.

Sympatric speciation of anole lizards has been observed on the above four islands of Puerto Rico, Cuba, Hispaniola, and Jamaica (Losos 2001, Losos and Schluter 2000). Each of the islands has species adapted to a variety of habitats. There are about 110 species of these lizards on those islands that have apparently evolved to fill most conceivable niches, such as living on tree canopies or on tree trunks, and feeding on twigs or on grass.

How do we know these species evolved sympatrically (together in the same locality) rather than allopatrically (each in a separate locality and then came together)? There are two strong indications of sympatric speciation among several of the species. Firstly, in the 1970s the brown anole, *Anolis sagrei*, was introduced to twenty islands, each of which offered them a variety of habitats. About twenty years later, on many of the islands, lizards of different morphologies were found that were adapted to the various habitats and were reproductively isolated from one another. Since, after twenty years, each island was found to have several adapted species, one can conclude that in the time between the introduction of the lizards to an island and the time the population was observed, the new species on each island diverged from the original *Anolis sagrei*.

Secondly, DNA analysis indicates that the various species on an island evolved on that island and did not immigrate from elsewhere. Environmental inputs may have induced genetic modifications as hypothesized in Chapter 3. How is this shown? The anoles on each of the four islands have the same types of adaptations. Species on different islands adapted to the same habitats have nearly the same morphology. For example, all four islands have a species of anoles adapted to living at the base of a tree and one adapted to the tree canopy. Three of the islands have species adapted to the tree trunks and a species adapted to living in grass. If these various species evolved their separate adaptations together in one place and then dispersed, one would expect the DNA of a lizard on one island to match more closely the DNA of a lizard having the same adaptation on another island than to match a differently adapted anole on the same island. DNA analysis, however, has shown the reverse — the variously adapted species on any one island are more closely related to each other than each is to similarly adapted species on the other islands (Losos 2001, Losos et al. 1998, Jackman et al. 1997). That is to say the DNA of the tree-top lizards is a better match to the tree-base lizards of the same island than it is to the tree-top lizards of another island. The implication is then strong that the first lizard population to colonize each island diverged independently to fill the available niches. Neo-Darwinian theory is at a loss to account for this type of rapid parallel evolution through DNA copying errors and natural selection, but it can be accounted for by the NREH. Evolution through a built-in response to an environmental input does not generate new information. The built-in mechanism was already in the lizards but it was latent. This is another example of evolution-in-action, which does not involve an increase of information. It gives no indication of how information could be built up and therefore lends no support for Common Descent.

* * *

Evolution of Darwin's Finches

On his famous trip on the *Beagle,* Charles Darwin visited the Galápagos Islands. Among the fauna he saw on the islands were birds that looked like finches but were not the same as any finches he had seen before. Expert study of his collected specimens back in London confirmed them to be new species of finches, unknown elsewhere. Darwin theorized that some time in the past, a few finches found their way to the Galápagos Islands from the mainland. He suggested that since then, variations (now assumed by neo-Darwinians to be DNA copying errors) have appeared in the birds, and these changes were subject to natural selection. As a result of this scenario, the birds are supposed to have diversified into the fourteen species now found on the islands.

Darwin theorized that the ancestors of the Galápagos finches flew, or were swept by winds, to the islands from South America, more than 600 miles distant, and biologists today agree (Grant 1986). DNA analysis indicates the likely candidate for the Galápagos-finch ancestor is a member of the grassquit finch of the species *Tiaris obscurus* (Sato et al. 2001) of South America. The diversification of the Galápagos finches is estimated to have taken about 2.3 million years (Sato et al. 2001).

Did the finches really take a few million years to diversify? We can speculate, but we don't really know when the first finches found their way to the islands. That is a matter of history and we have no historical records that can confirm that speculation. Darwin's suggestion, and today's conventional wisdom on the subject, is no more than speculation. But in a controlled study, finches were introduced to an island that previously had no finches (Conant 1988, Pimm 1988). In 1967, about 100 identical finches were removed from a U.S. Government Bird Reservation in the middle of the Pacific Ocean and were taken about 300 miles away to a group of four small atolls lying within less than ten miles of each other, which had no native finches. The birds were released onto one of these islands, and they soon spread to all of them. Seventeen years later, when the birds were first checked, they were found to have a variety of bill shapes and to be adapted — both by their behavior and by their bill shapes and associated muscles — to various niches. This was a speeded-up form of the conventional scenario of Galápagos finch evolution. In seventeen years, and possibly less, the finches had diversified into various niches. If this diversification occurred in less than seventeen years, why did Darwin's Galapágos finches have to take two million years? They could have done it much more rapidly, and perhaps they indeed did. The diversification can be accounted for by a built-in response of the finch's genome to an environmental input as postulated in the NREH.

Each species of finch is adapted to its own niche, with its beak shape, muscles, behavior, and other phenotypic characters appropriate to its niche. The proximate biochemical signal evoking the change in beak shape has been discovered to be a protein growth factor called *Bmp4* (*bone morphogenic protein 4*). During embryonic development of the finch (and very likely of other birds as well [Wu et al. 2004]), the more *Bmp4* that is made, the broader and deeper is the bird's beak (Abzhanov et al. 2004). This protein acts as a signal to the development of the craniofacial bones, which, among other things, determines the beak's shape. If my suggestion is correct that the hormones triggered by environmental inputs affect embryonic development, then those hormones induce these growth factors to form the finch beak. It is likely that the environmentally-induced hormones stimulate other growth factors as well. The built-in mechanism of the NREH enables the bird population to adapt to a new environment quickly and efficiently without having to call upon the slow and wasteful neo-Darwinian process of random mutation and natural selection. An evolutionary adaptation that would take millions of years waiting for the right DNA copying errors and natural selection can be done in a single generation through the mechanism of the NREH. The evolution of the finches, both those of 1967 and those Darwin saw and reported, gives no indication of a buildup of information and therefore offer no support for Common Descent.

* * *

Character displacement is a speciation phenomenon in which two species that tend to be similar when they live apart will diverge to become different in one or more characteristics when the two species occupy the same territory (Brown and Wilson 1956). This capability is apparently designed to help reduce the competition between species. The differences that develop between them are generally thought to be genetically-based. A striking example of rapid character displacement in which the divergence appeared after two similar species resided together for only twenty-two years, and in which the character displacement occurred rapidly — in just one year — was reported by Peter and Rosemary Grant (Grant and Grant 2006). The medium ground finch (*Geospiza fortis*) was the sole finch species on an undisturbed Galápagos island until 1982 when two females and three males of the large ground finch (*Geospiza magnirostris*) arrived. As these new arrivals bred, the size of the beaks of the *G. fortis* were monitored. The mean beak size remained constant[1] until 2004, when it suddenly underwent a substantial decrease in a single year.[2] The change was adaptive in that it permitted the *G. fortis* to feed on small seeds, avoiding competition with the larger and now well-established *G. magnirostris*. In this example, the environmental input was the competition from the existing species. The competitive stimulus built up slowly, starting from 1982. Until 1997 the number of *G. magnirostris* was too small to estimate. From that time they built up gradually, and by 2004 they were suddenly almost as numerous as the *G. fortis*, and that is when the abrupt change in mean beak size of the latter underwent its sudden change. Here the evolutionary phenomenon surely points toward a nonrandom genetic change triggered by an environmental input. It certainly does not lend any support for the neo-Darwinian theory of random mutations and natural selection.

Other examples of character-displacement evolution occurring rapidly are the body sizes of solitary and paired lizards on small islands (Schoener 1970, Losos 1990), the body size of mud snails (Fenchel 1975), and mandible lengths of tiger beetles (Pearson 1980). These examples

---

1 Except for a temporary increase, which started before the arrival of *G. magnirostris*.
2 The decrease was four times the one-sided 95% confidence limits on the estimate of the mean in 1973.

of rapid evolution give no indication of how genetic information might be increased. They therefore offer no support for Common Descent.

* * *

There are many examples of rapid evolution in the literature beyond what I have given here. These include the evolution of the Hawaiian mosquito fish: its age and length at maturity, growth rates, and size of offspring (Stearns 1983); the evolution of the soapberry bug: its beak length and development (Carroll et al. 1997); and the evolution of morphological traits in grasses (Snaydon and Davies 1972). Thompson (1998) gives a list of more than twenty examples of rapid evolution observed today. None of the evolution in these examples shows how information can build up and therefore none of them offers any support for Common Descent. In all of the above examples there is not enough time for random mutations to occur or for natural selection to work. The only reasonable alternatives for a mechanism of how they work are that either the adaptive genotypes were already in the population or they became manifest through an environmental input as described by the NREH. Neither gives any support for Common Descent.

* * *

Microevolution and Macroevolution

The terms *microevolution* and *macroevolution* are used to distinguish small evolutionary changes from large ones. Microevolutionary changes are heritable changes that can be observed in the laboratory or in nature, and the changes are usually small. Macroevolutionary changes are not directly observable but are inferred from fossil or molecular data. Macroevolution is said to take long spans of time and to lead to new categories, usually above the species level, leading to new orders or phyla.[3] This type of evolution cannot be observed but can only be inferred because they are said to require times periods far exceeding the human lifespan.

---

[3] These categories are defined in my previous book, *Not By Chance!*

At the beginning of the twentieth century, there was some controversy among Darwinists as to whether or not macroevolution was qualitatively different from microevolution (Hull 1970). Some felt large evolutionary changes were the result of sudden large changes in body form and had nothing to do with the natural selection of Darwin's theory. The suggested process producing these large evolutionary changes, independent of natural selection, was called *orthogenesis*. But by the end of the twentieth century, the consensus among evolutionary biologists was wholly against orthogenesis.

The term *macroevolution* was introduced in 1927 by the Russian geneticist Yuri Filipchenko (1882-1930) in a book he wrote in German entitled *Variabilität und Variation* (*Variability and Variation*) (Erwin 2004). He maintained that small Darwinian changes could produce evolution up through the appearance of new species, but evolution above that level had to involve different processes. Theodosius Dobzhansky (1900-1975), a renowned geneticist and evolutionary biologist who had been a student of Filipchenko in Russia and who brought Filipchenko's term *macroevolution* to the English-speaking biologists, disagreed with his mentor and held that macroevolution was nothing more than an accumulation of a long succession of microevolutionary events. He wrote:

> ... there is no way toward an understanding of the mechanisms of macroevolution, which require time on a geological scale, other than through a full comprehension of the microevolutionary processes. For this reason we are compelled at the present level of knowledge reluctantly to put a sign of equality between the mechanisms of macro- and microevolution. (Dobzhansky 1937)

Emmet Reid Dunn (1894-1956), a leading herpetologist, agreed with Dobzhansky about macroevolution. He wrote:

> No evidence of a distinction between microevolution and macroevolution is evident (Dunn 1943).

Everett C. Olson (1910-1993), a leading vertebrate paleontologist, wrote that the prevailing opinion was that macroevolution was nothing more than a long sequence of microevolutionary changes, and that no new principles were involved. Both microevolution and macroevolution were products of random mutation and natural selection (Olson 1965).

The present consensus among Darwinists on the issue is that macroevolution is merely an extension of microevolution. The arguments presented here, however, offer a different resolution. In reality, though, the issue is moot because <u>there is no evidence for macroevolution</u>. There is evidence only for microevolution.

* * *

Punctuated Equilibrium

In 1972, invertebrate paleontologists Niles Eldredge and Stephen Jay Gould (1941–2002) proposed a theory they called *punctuated equilibrium* to account for the macroevolutionary changes interpreted from the fossil record (Eldredge and Gould 1972). They opposed the prevalent view that the lack of intermediate forms in the fossil record must be explained as an imperfection in the record. Indeed, the major thrust of their theory was to emphasize that the absence of fossil data for gradual evolution is real. They held that the fossil record predominantly showed long periods of little or no change (stasis), <u>punctuated</u> by <u>short bursts of rapid evolution</u>.

These paleontologists argued that the ubiquitous examples of stasis and punctuation in the fossil record must be accounted for and are not to be dismissed as imperfection in the record, as Darwin had done and as had been the custom of most Darwinists since. All Darwinists agree that large populations do not easily undergo evolutionary change, particularly when the environment is stable. This stability can explain the many examples of stasis in the fossil record.

Eldredge and Gould explain examples of abrupt appearances of new species as the result of allopatric speciation, which they describe in the following way. A small portion of a population on its periphery separates from the main population. This small population will most likely have an unrepresentative sample of the genetic makeup of the parent population, causing it to evolve away from the parent (called the "founder" effect), and if its isolation lasts for a sufficiently long time, it may evolve into a different species. If the new species then should return to its ancestral home, it will no longer be able to mate with its parent species because its reproductive characteristics would have evolved away from those of its ancestors (a process called *reproductive isolation*). For example, its chromosomes may have evolved to a point where hybrids would not be viable, or it may have developed new mating preferences. Thus there will be no gene flow between the new species and the ancestral one. If the new species should then have an adaptive advantage over the old species, the new species may replace the old one through natural selection and the fossil record of this process will show a long period of stasis of the ancestral species followed by an abrupt (in geologic time) transition to the new one.

They suggested that there might be more than one peripheral offshoot of the ancestral population, which might then return after having evolved into a new species, and these returning offshoots would compete with each other in taking over the population. The punctuated-equilibrium theory thus shifts the emphasis from competition within a single species to competition among species. Of course, the same old Darwinian process of random mutation and natural selection operates within the peripheral populations but the punctuationists emphasize the selection among species (*species selection*) as a new feature, which, they say, goes somewhat beyond Darwinian theory.

Eldredge and Gould have made an important contribution in pointing out that the fossil record really shows long periods of stasis punctuated by relatively short bursts of speciation. It seems to me, however, that their suggestion of allopatric speciation is a little off the mark. I suggest that the "short bursts of speciation" more likely stem from sympatric speciation of the kind exemplified in the several examples I have given above. Some of these examples are representative of many occurrences of environmentally-induced genetic and phenotypic changes that we have observed to occur in very short times — even within a single generation. The model of Eldredge and Gould requires a portion of the population to wander off, then evolve into a new species reproductively isolated from the parent population, then wander back, mix with the parent, and finally take it over. Actually, they speculate that this happens several times and that the returning new species compete for the takeover. My suggestion is simpler in that it requires no wandering off and coming back. It relies only on the well-established mechanism of an environmental stimulus triggering an adaptive genetic change as described in Chapter 2. In such a case, species change can occur rapidly — not just rapid with respect to geologic time, but rapid even in terms of human time. When Eldredge and Gould published their papers on punctuated equilibrium, sympatric speciation was still thought to be highly unlikely, so they were naturally drawn to advocate allopatric speciation as the dominant process following the opinion of Ernst Mayr, who was known as the doyen of twentieth-century biology (Hölldobler 2004). Since then, as I have noted above, Mayr had come to favor sympatric speciation following the abundance of data in its favor. In accord with what we know now and from my discussion of the NREH above, sympatric speciation under the NREH is a more likely explanation of the punctuation phenomenon.

* * *

The kind of rapid evolution described in this chapter is the only evolution we can witness and is the best evidence we have for evolution. The kind of evolution in the examples of rapid evolution, however, produces no increase of information in the genome and therefore offers no clue as to how evolution can increase information to evolve today's complex life from some alleged original primitive cell. Advocates of Common Descent have not adequately addressed the issue of how the information in complex present-day organisms was built up from the alleged remote common ancestor of life. As I have stated, without an explanation of how evolution increases information, there is no theory to account for Common Descent.

Chapter 4
The False Arguments for Evolution

Here is a typical Darwinist claim:

"What evolution has is what any good scientific claim has — evidence, and lots of it. Evolution is supported by a wide range of observations throughout the fields of genetics, anatomy, ecology, animal behavior, paleontology, and others. If you wish to challenge the theory of evolution, you must address that evidence. You must show that the evidence is either wrong or irrelevant or that it fits another theory better. Of course, to do this, you must know both the theory and the evidence."

(www.talkorigins.org/faqs/faq-misconceptions.html)

In this chapter I examine the evidence for Common Descent, colloquially known as *evolution*. The modern professional literature does not seem to deal much with the evidence. Perhaps that is because they consider the truth of Common Descent to have been settled, making a systematic discussion of the evidence superfluous. Perusal of the professional literature turns up examples of new discoveries that are said to be evidence of Common Descent, but general discussions of the evidence are few. Systematic discussion of evidence for Common Descent is best found in books written to convince laymen of the truth of evolution. Perhaps the best of these are the relatively recent publication of the National Academy of Science (NAS 2008) and three later books (Coyne 2009, Dawkins 2010, Rogers 2011).

The evidence presented includes fossils, the phylogenetic tree, biogeographical data, vestigial organs, and examples of evolution in action. Using the background I have laid down in the preceding chapters, I will show why all these arguments for Common Descent are flawed. I will also present an alternative explanation of the evidence that I think is better than that of Common Descent. It turns out that there is no good evidence for Common Descent.

As I have noted earlier in this book, a distinguishing feature of life is the information it contains and its ability to use that information to function and reproduce. The major challenge to Common Descent is to account for how that information has been built up through a natural process. After all, contemporary organisms such as fishes, elephants, and people obviously contain much more information than could have been in the alleged primitive cell from which all life is claimed to have evolved. It has not been shown how a natural process could build this information, which must be done if we are expected to consider Common Descent a proper scientific theory.

Can neo-Darwinian theory account for that buildup of information? Have Darwinists ever shown how the information in living organisms could have been built up by the mechanism of random mutations and natural selection? Does any of the evidence usually cited relate to the buildup of that information? The answer to all these questions is: No.

* * *

The Phylogenetic Tree as Evidence for Common Descent

If the Common-Descent paradigm were true, one should be able to construct a tree of organisms that shows the pattern of decent. A phylogenetic tree, sometimes called "the tree of life," is a diagram showing what is thought to be the evolutionary descent of animals and plants. A tree can be constructed by following a particular set of organs through different kinds of animals (or plants). One could, for example, group animals according to the similarity of their limbs, or of their digestive system, or of their blood-circulatory system. Using limbs as a characteristic, for example, the grouping would take the form of a tree where animals with most similar limbs would be placed on the same tree branch, while those with less similar limbs would be placed on different branches. The way branches ramify is intended to indicate how the animals evolved one from the other. Alternatively, one could group animals according to some other feature, such as the blood-circulatory system. If trees constructed by comparing several different organs or physiological systems are identical, then one would have some confidence that the tree is meaningful. If comparing all possible biological features yields the same tree, then the tree could have some objective reality. Richard Dawkins (2009, pp. 321 ff.) offered what he calls "powerful evidence" for Common Descent based on the (presumed) existence of a consistent phylogenetic tree.

According to the neo-Darwinian paradigm, animals that have similar characteristics should be on the same branch of the tree. All animals that share a characteristic should also share a latest common ancestor, which would be the first one to have that characteristic. Dawkins contends (and as far as I know all Darwinists agree) that the ability to arrange all living organisms in a unique tree-like chart would be a proof of Common Descent.

The logic here, however, is backwards. Logic would say that if Common Descent were true, one should be able to construct a unique phylogenetic tree. They have reversed the logic. The best one can say is that a unique phylogenetic tree would be consistent with Common Descent. But that doesn't prove Common Descent any more than it proves Creation. After all, a good phylogenetic tree is also consistent with Creation. Living organisms were first put into a logical ordering by the Swedish naturalist Carolus Linneus (1707-1778), who worked under the assumption of Creation as did all scientists of his day. His system of classification is still in use today. Dawkins dismisses, without explanation, the correspondence between the classification system and a Creation hypothesis as "far-fetched." As we shall see, however, the correspondence with Common Descent is more deserving of that epithet.

Because the ability to construct a unique tree is a required consequence of Common Descent, the inability to construct such a tree should be a contradiction to it. As we shall see, there is no unique tree, although Darwinists have attempted to explain away the contradiction. I shall describe these efforts shortly.

Although phylogenetic trees have traditionally been constructed by comparing anatomical features, in the last few decades it has become popular to do it by comparing the amino-acid sequences of proteins or the nucleotide sequences of DNA. Organisms sharing a common protein (which is typically a chain of a few hundred amino acids), for example, would be assumed to have a latest common ancestor that was the first to have that protein. Dawkins's argument as to why phylogenetic trees are evidence for evolution is that such a tree is "not what you would expect if a designer (i.e., Creator) had surveyed the whole animal kingdom and picked and chosen — or 'borrowed' — the best proteins for the job." Note the absurdity of Dawkins, the arch-atheist, using a theological argument to support the dogma of evolution. However, his argument for evolution fails not only because it rests on amateur theology but also because it is factually wrong.

An argument for Common Descent would be helped if anatomical data and molecular data would always lead to the same tree. However, the fact is they don't (Heled and Drummond 2010, Rosenberg and Degnan 2010, Degnan and Rosenberg 2009, Degnan and Rosenberg 2006, Nichols 2001). Phylogenetic trees based on different genes are known to give contradictory results. There was hope that the use of whole genomes, or at least large portions of genomes, for phylogenetic studies would resolve those contradictions, but that only made the problem worse (Jeffroy et al. 2006, Dalos et al. 2012).

The lack of uniqueness of the phylogenetic tree is usually explained away by what is called "convergent evolution." <u>*Convergent evolution* is the appearance of the same trait or character in independent lineages.</u> <u>It is, however, an invention.</u> It was invented solely to avoid addressing the failure of the phylogenetic tree to support Common Descent. There is no theoretical support for convergence, and whatever evidence has been given for it is the product of a circular argument. Richard Dawkins (2010) seems to revel in describing numerous examples of convergent evolution without realizing that any one of those examples destroys his case for evolution.

Convergences have been invoked throughout the animal and plant kingdoms. A fascinating example of "convergence" has been recently reported between the auditory system of mammals and that of insects (Montealegre-Z et al. 2012). The vertebrate ear converts sound waves to electrical nerve signals, which go to the brain. Sound waves in the air enter the ear, travel through the auditory canal, and impinge on the eardrum (tympanal membrane), causing it to vibrate. The eardrum is connected to a lever system of three bones (hammer, anvil, and stirrup) in the middle ear. This lever system transmits the vibrations of the eardrum to the fluid in the cochlea. The apparent complexity of the concatenation of these bones in the ear is designed to match the impedance of a sound wave in the air to its impedance in the cochlear fluid, thus reducing reflections from the interface and the consequent loss of signal. The vibrations in the fluid are in turn transmitted to an array of acoustic-detector hair cells immersed in the fluid, which resolve the sound signal into its frequency components. The vibrations of the hair cells are converted to electrical nerve signals by the cells attached to each hair and these signals are transmitted to the brain.

The South American rainforest katydid *Copiphora gorgonensis* has been discovered to have a hearing system analogous to that of the vertebrates. It has:

- a tympanal membrane analogous to the eardrum,
- a system of levers to transmit the vibrations to a fluid and

for impedance matching, analogous to the three bones in the inner ear of a vertebrate, and
- an acoustic vesicle filled with fluid, analogous to the cochlea, which transmits the sound waves to a frequency-analyzing device that sends the analyzed signal on to the brain.

The authors call this "a notable case of convergence" where mammals and katydids "have evolved to hear in a markedly analogous way." But neo-Darwinian theory cannot account for this "convergence."

I have shown (Spetner 1997) that evolution itself cannot work according to the neo-Darwinian paradigm unless at every stage of evolution there are a huge number of potentially beneficial mutations. There is no hope of the neo-Darwinian mechanism working unless there are many potentially adaptive mutations at each stage on which chance could work (which no one has shown there can be). But if there are, then clearly there can be no convergent evolution because the number of possible evolutionary paths from any stage would be extremely large, making a repetition of a previous path highly unlikely and neo-Darwinian evolution therefore unpredictable.[1] If there are enough potentially adaptive mutations at each stage to allow neo-Darwinian evolution to work, convergent evolution must be impossible.

The inherent contradiction between convergent evolution and the ability of neo-Darwinian evolution to work at all puts evolutionary theory between a rock and a hard place. I think these considerations of neo-Darwinian theory may be what inspired the late Stephen Gould (1989) to state, in regard to his hypothetical experiment of "replaying life's tape,"

> I believe that ... any replay of the tape would lead evolution down a pathway radically different from the road actually taken.

Because the nature of neo-Darwinian theory requires that any "replay

---

[1] There is a controversy among Darwinists as to whether or not and to what extent adaptive evolution is predictable (Wood et al. 2005).

of the tape" will not yield the same result a second time, any construction of the tree of life should also not have similar characteristics showing up independently on separate branches. There should be no convergent evolution. According to the theory, all sets of common characteristics should each be traceable to a unique ancestor.

But it turns out phylogenetic trees do not fulfill this condition. Although Gould's view is based firmly on the neo-Darwinian paradigm, when his view was tested experimentally, it was found to fail (Travisano et al. 1995, Losos et al. 1998, Blount et al. 2008). This, by itself, is evidence of the failure of neo-Darwinian theory. There are many examples of spoilers in the classification system of living organisms that violate the condition that there should be no examples of organisms having similar complex characteristics not traceable to a unique ancestor. Because of the contradiction between the Darwinian paradigm and actual observations and experiments, it would be reasonable for Darwinists to have abandoned the paradigm, or at least to have questioned it, or at the very least to have stopped using the phylogenetic tree as evidence for Common Descent. Nevertheless, Darwinists invoke *convergent evolution* to explain away the discrepancy between theory and experiment, even though they cannot show that it works. L. L. Parker et al. (2013) wrote, "Convergence is not a rare process restricted to several loci but is instead widespread." They are compelled to invoke convergence in almost all of their discussions of evolution.

The neo-Darwinist invention of convergent evolution reminds me of a story about a poker game on a Mississippi riverboat, as told to me by the late Sam Goldberg of St. Louis:

Three professional St. Louis gamblers were engaged in a high-stakes game and a New York card shark asked to join. Well, the New Yorker had very agile hands and he could make a deck of cards do anything he wanted. It didn't take long before he had dealt himself a royal flush. The bidding went up and up and when the only other player left had called, he opened his hand and went to take his winnings.

Then his opponent said, "Wait a minute!" and opened his hand showing 2-4-6-8-10.

"What's that?" asked the New Yorker, puzzled.

"That's a LOLLAPALOOZA! And on the Mississippi River, the rule is a lollapalooza beats anything!"

The New Yorker had never heard that one before, but he was up to the challenge and before long he dealt himself a lollapalooza.

When he went to take his winnings, the others said,

"Hold on there! The rule on the Mississippi River is only one lollapalooza is allowed per day!"

Convergent evolution is the Darwinists' lollapalooza. They made it up to keep their phylogenetic tree from falling apart, but they can't say how convergence happens. As Joseph Keating (2002) wrote in another context, it is no more than a "pseudo-explanation, and may deceive us into believing we have explained some aspect of biology when in fact we have only labeled our ignorance." Darwinists haven't made a probability analysis to see if convergence works through random mutations and natural selection, just as they haven't made that analysis for evolution in general. In no other branch of science can one get away with inventing a new concept just to solve a difficulty without showing, either by calculation or other competent examination, to what extent it is reasonable. It happens to turn out that convergence is not reasonable.

Other interesting examples of "convergences" that cannot be accounted for by current evolutionary theory are:

- Whales, dolphins, and bats are the only mammals that have echolocation systems. These systems enable the animal to "see" by means of sound waves, as in naval sonar systems. The animal transmits sound signals and locates objects in its environment by analyzing the return signal or echo. To "see" with the echolocation system, the animal must transmit strong sound signals and be able to detect weak reflections from objects in its surroundings. The animal's brain must be wired to interpret these signals much as our brains interpret the visual inputs from our eyes. On the basis of gross anatomy, the whales and dolphins are classified closer to the cow and other ruminants than to the bats. But on the basis of their echolocation systems, they should belong with the bats. Liu et al. (2010) have stated, "The ability of some bats and all toothed whales to produce sonar pulses and process the returning echoes for prey detection and orientation (echolocation) is a spectacular example of phenotypic convergence in mammals." Not only that, but a molecular analysis of the protein prestin puts whales and dolphins squarely with the bats. Prestin is a sound-sensitive protein that is essential for hearing in general, and in particular it is part of the hearing mechanism of the echolocation system. The prestin molecules in whales and dolphins have fourteen amino acids that are not in the prestin molecule of any other mammals except for the bats, whose prestin protein also has those same fourteen amino acids (Liu et al. 2010). This is strong evidence that the phylogenetic tree does not represent objective reality and should cast a pall over Common Descent. But Darwinists rescue Common Descent by declaring that the prestin protein and the entire echolocation systems in whales, dolphins, and bats are the results of convergent evolution. But, as I have pointed out above, they have never shown that convergent evolution is anything more than their lollapalooza.

- Even among the bats alone, convergence had to be invoked. Both the mustached bat and the horseshoe bat have an echolocation system and are thought to have developed them independently. Neuweiler (2003), apparently trying to rescue Common Descent, called this one of the most striking examples of convergent evolution.

- Both insects and mammals detect chemical cues in the environment through odor receptors on the surfaces of cells in the olfactory sensory epithelium. The olfactory systems of both insects and mammals contain a large number of odorant-receptor genes — about 60 in *Drosophila* and about 1,000 in mice (Thorne et al. 2004). These receptors, in combination, enable the animals to discriminate among hundreds of thousands of different odors (Buck and Axel 1991). The brain receives the combined neurological signals from these cells and interprets the corresponding odor.
- The taste cells in the tongue of both insects and mammals are dedicated to the same individual basic taste sensations (sweet, bitter, etc.) and they have about the same number of taste receptors (Thorne et al. 2004). The strong similarities between the olfactory and taste systems in mammals and insects should be an embarrassment for Common Descent, but the Darwinists try to rescue it with their invention of convergent evolution (Thorne 2004, Morris 2009).
- The eye is said to have evolved independently at least forty times and probably as many as sixty-five times (Salvini-Plawen and Mayr 1977, as cited by Land and Fernald 1992).[2] To rescue Common Descent, Darwinists invoke convergence, even though they have no idea how it can work.
- Bossuyt and Milinkovich (2000) reported that their analysis of nuclear and mitochondrial DNA shows burrowing frogs from Madagascar and frogs from India evolved independently. Here, too, the authors try to save Common Descent by using their lollapalooza to claim these frogs have "converged" in their morphological, physiological, and developmental characters.

---

[2] The notion of the eye evolving even once is what Darwin said gave him a "cold shudder."

- Brown and Kodric-Brown (1979) studied the pollination ecology of nine species of red tubular flowers that bloom together in different combinations in the White Mountains of Arizona. They wrote, "All species were strikingly *convergent* in floral color, size, and shape." (My emphasis) Rather than abandon Common Descent, they choose to claim convergence even though they don't know how that could have possibly happened.
- Five species of recently extinct, nectar-feeding songbirds that had been called "Hawaiian honeyeaters" had long been classified with the Australasian honeyeaters (*Meliphagidae*), because they looked and acted like them. Fleischer et al. (2008) analyzed the DNA of museum specimens of these extinct birds and concluded that they were not meliphagids. Their striking similar characteristics led the authors, in an apparent attempt to save Common Descent, to report that these two groups of birds are a "particularly striking example" of convergent evolution.
- Nectar-feeding insects such as the hawkmoth are said to have converged with the hummingbirds in size, wing-beat frequency, and metabolic rate, as well as in their morphological, physiological, and biochemical characteristics (Welch et al. 2006). Hummingbirds and insects also produce lift during hovering flight by using similar aerodynamic mechanisms (Warrick et al. 2005). This similarity refutes Common Descent, which the authors try to retain by invoking convergence. Moreover, bats are also said to have converged in their hovering aerodynamics to those of hummingbirds and insects (Dickinson 2008).
- Convergence is invoked as an apologetic for evolutionary similarities among the proteins in the venoms found in all animal phyla, including arthropods, cephalopods, and vertebrates (Fry et al. 2009).

- Feather lice in birds are morphologically different from body lice and head lice in the same birds. Feather lice in different species of birds are, on the other hand, morphologically very similar. On the basis of morphology alone, it had been supposed that feather lice evolved from more general lice and then spread to various species of birds. But a recent study (Johnson et al. 2012) has shown that the different types of lice on any single bird species are genetically more similar to each other than the same type of lice on different bird species. This indicates that the lice diverged into feather lice and body lice on each bird species separately. The phylogenetic tree for these lice based on morphology is completely different from the tree based on genetics, making it impossible for Common Descent to stand. Of course, the authors try to relieve the embarrassment by claiming the similarities to be the result of convergent evolution.
- The milkweed is a family of poisonous plants containing cardenolides, which protect the plant from being eaten by animals. The plant is toxic to animals because the cardenolides block the animal's sodium-pump molecule (ATPase), which is essential for muscle contraction and other neural functions. However, some animals, such as the monarch butterfly and the leaf beetle, are unaffected by the cardenolides and actually feed on the plant. Moreover, they store the cardenolides in their bodies, making themselves poisonous to their own would-be predators. Their ATPase molecules are insensitive to the cardenolides because the amino-acid sequence of their proteins differs from that of other animals. The ATPase of the monarch butterfly and of the leaf beetle have the same amino-acid difference and they are said to have arrived at the same ATPase molecule by convergent evolution (Dobler et al. 2012, Zhen et al. 2012). Fourteen species of insects have been

shown to share the insensitivity to the cardenolides. All are said to have independently evolved the same mechanism that gives them the insensitivity. One researcher expressed recognition of the difficulty of explaining these data as the result of random mutations. He commented, "The finding of parallel evolution in not two, but numerous herbivorous insects increases the significance of the study because such frequent parallelism is extremely unlikely to have happened simply by chance" (Princeton University 2012).

The list can go on and on of the instances where Darwinists are compelled to invoke convergence to rescue Common Descent. Each example attributed to convergence is an expression of ignorance as to why animals that are classified far apart in the phylogenetic tree have very similar complex features. The dilemma here stems from the Darwinists' insistence on the Common-Descent paradigm.

Darwinists have another way of explaining away the lack of a unique phylogenetic tree besides convergent evolution. They postulate that plants and animals can transfer DNA fragments between species, a process called *horizontal gene transfer* (HGT).[3] HGT is known to occur among prokaryotes such as bacteria. Bacterial cells have no nucleus to quarantine off their DNA from the rest of the cell. They can eject DNA fragments from one cell to another either in a kind of sexual transmission or by one bacterium simply emitting a piece of DNA, in the form of a plasmid, into the external medium, which can be picked up by another bacterium. Much of the antibiotic resistance gained by pathogenic bacteria comes through HGT from resistant bacteria.

There is no direct evidence, however, of such transfer among different species of eukaryotes whose DNA is bound inside the nucleus of the cell. As Jan O. Andersson of Uppsala University has pointed out, whatever we know about gene transfer in eukaryotes is based on anecdotal evidence

---

3 Also known as lateral gene transfer (LGT).

(Andersson 2009). Moreover, according to Keeling and Palmer (2008), the examples of HGT are usually identified by discrepancies between the phylogeny deduced from anatomical data and that deduced from genetic data, so explaining the discrepancies in the phylogenetic trees by HGT is a circular argument. Syvanen (2012) has noted (without citing a reference) that HGT in eukaryotes has been observed in the laboratory but so far not in a natural environment. Thus we don't have any direct evidence yet of HGT in eukaryotes. Perhaps for this reason, Darwinists prefer to explain the phylogenetic-tree difficulties through convergence rather than through HGT. If ever HGT is verified to be an important player in the changing of populations, the entire business of phylogenetic trees would be thrown into chaos (Syvanen 2012). In any case, the state of phylogenetic trees today offers no support for Common Descent.

* * *

Transposons as Evidence for Evolution

Alan Rogers (2011) wrote that transposons[4] are used to estimate phylogeny, and that they testify to the truth of Common Descent. He argues that since transposons are not functional and insert themselves across the genome at random, they can be used to establish a phylogenetic tree. He wrote (p. 27):

> ... it is extremely unlikely that two transposons will ever insert at the same spot. This means that two individuals who share the same transposon must also share an ancestor.

It turns out, however, that his premise is false. First, transposons are functional (Emera and Wagner 2012) — not nonfunctional as Rogers wrote. As we have seen in Chapter 3, they may well be responsible for reorganizing the genome as an adaptive response to an environmental input. Second, they do not insert themselves at random. James Shapiro, who has been studying transposons for more than thirty-five years,

---

4 The "jumping genes" discussed earlier, which consist of integrated systems of proteins and nucleic acids.

reports that transposons insert themselves at preferred spots (Shapiro 2011). Other experts in transposons have also found that transposons insert at preferred sites (Levy et al. 2010). Consequently, two transposons in the same spot in the genome of two species is no indication that they share an ancestor. So much for Rogers's attempt to claim evidence for evolution from transposons.

* * *

The Fossil Evidence for Evolution

If I were to confine some rabbits in a cage, and some time later I find more rabbits in that cage, I can be certain that the earlier rabbits gave birth to the later ones. If fossil shells are found in geological strata, one above the other, and they all look more or less alike, I might reasonably suppose that the upper ones are the progeny of the lower ones, but I would not be quite so certain as with the rabbits. If the upper ones are somewhat different from the lower ones, I might still be inclined to assume that the lower ones gave birth to the upper ones, but I nevertheless have serious reservations about that assumption. Note that in such a case, the paleontologists Niles Eldredge and Stephen Gould (1972) have suggested that the upper ones were often not born in that place, but have moved into the territory from somewhere else. So the assumption that fossils in the upper layers descended from those in the lower layers is not necessarily the only interpretation of fossil data.

Darwinists seem to believe that fossils are proof of Common Descent. Just as the logic used for the phylogenetic trees is backwards, so, too, the logic used for the fossils is backwards. If Common Descent were true, then one would expect to find fossils showing the stages of evolution. But the logic does not work the other way around. The most one can say is that the fossils would be consistent with an evolutionary story. I call it a story because there is no evolutionary "theory" that can account for Common Descent. Gertrude Himmelfarb has put it well when she said:

> Every paleontological discovery that seems to have evolutionary significance is somehow taken as confirmation of the theory of natural selection, even when it has not the remotest bearing upon that theory. (Himmelfarb 1962, 446)

The fossils can, however, be accounted for by the NREH, independently of Common Descent, and shortly I shall give an example.

Why don't the fossils that show small changes from one layer to the next support Common Descent? They don't because the changes may well be the kind of changes described in the examples of the previous chapter. Just as those examples did not involve any gain of genetic information (remember, either the genetic changes were the result of a built-in program or the "new" genes were already in the population), so the fossil evidence may not be evidence of a gain of genetic information. In neither case is information added to the biosphere so those "evolutionary" changes cannot lend any support for Common Descent. A theory of Common Descent, I remind you, must account for the buildup of information.

The arguments the Darwinist authors give on the basis of the fossil record are, as with the phylogenetic tree, theological rather than scientific. Their argument is that the Creator would not have put fossils in the ground to confuse us. On this basis they conclude that there was no Creation and therefore Common Descent must be true. Those authors are not theologians. Yet they have the temerity to offer theological arguments. I am not a theologian either, and I do not intend to make theological arguments. I am offering only scientific arguments about the lack of evidence for Common Descent. Nevertheless, I cannot refrain from showing the naiveté of their theology. I cite a theologian of two centuries ago, Rabbi Israel Lifshitz, an outstanding Jewish scholar of his time, who held ideas about fossils rather different from those of Dawkins. Rabbi Lifshitz gave a discourse in 1842 that included references to recent fossil finds. He did not think that the animals who left those fossils were the ancestors of living animals. He instead cited the Talmud and ancient Kabbalistic texts and said:

דאמרי' בב"ר ויהי ערב ויהי בוקר (דק' וכי מאחר שלא היה עדיין שמש בעולם, ערב ובוקר מניין) אמר ר' אבוה מכאן שהיה סדר זמנים קודם לזה וכו', מלמד שהי' הקב"ה בונה עולמות ומחריבן. ... וסוד ה' ליראיו, שנמסר להם שאנחנו כעת בהקפה הד'.

[As it is said in *Bereshit Rabbah*, "And it was evening and it was morning" (which raises the question: Since (on the first day) there was not yet a sun, how could there be evening and morning?). Rabbi Abuha said: From here it is evident that there was a time order before this, etc., to teach that the Holy One was building worlds and destroying them. ... A secret of G-d given to those who fear Him, as it has been passed down to them that we are presently in the fourth epoch.] (My translation)[5]

He then went on to describe several current fossil finds of animals that do not exist today, and commented:

מכל האמור נראה ברור שכל מה שמסרו לנו המקובלים זה כמה מאות שנים, שכבר היה עולם פ"א ושוב נחרב וחזר ונתקומם זה ארבע פעמים, ובכל פעם העולם התגלה בשלמות יתירה יותר מבתחלה, הכל התברר עכשיו בזמננו באמת וצדק.

[From all that has been said, it is apparent that everything handed down to us by the Kabbalists several hundreds of years ago has now become crystal clear in our time: that there was already a world and it was repeatedly destroyed and restored four times, and each time the world was revealed in greater perfection than before.] (My translation)

He did not see the fossils as a challenge to the Torah (Bible), but rather as a confirmation of it. His comment on the fossil finds was:

ועתה אחי ידידי ראו על איזה בסיס אדני תה"ק מונחים, כי הסוד הזה שנמסר לאבותינו ורבותינו, והם גלוהו לנו זה כמה מאות שנים מצאנוהו שוב בהטבע ברורה עיניו בזמנים המאוחרים כבזמנינו הבהירה ביותר.

[And now, my friends, see on what basis rest the foundations of our Holy Torah. Because this secret, which had been handed down to our forefathers and sages, and which they revealed to us many centuries ago, we have lately rediscovered in nature in the clearest manner.] (My translation)

---

5 The substance of this discourse is printed (in Hebrew) in the *Yachin uBoaz* edition of the Mishnah at the end of Order Nezikin, vol. 1.

I shall not pursue this matter further because I want to concentrate on the science. I bring this theological point only as a contrast with the naïve theological arguments of the Darwinists.

The Darwinist authors cite the circumstantial evidence of fossils of what they call "transitional" forms as proof of Common Descent.[6] Of course, as I have noted, this evidence proves nothing of the kind because it does not have a theory to cling to. Similar arguments from the fossil record could be given for Creation. This kind of "proof" is not scientific.

Darwinists cite fossils that they declare to be intermediate between major groups claiming they show that evolution has occurred. They maintain that these fossils are the remains of animals that were on an evolutionary transition from one major form to another. The example of a claimed transition from fish to the four-legged land-living animals (called *tetrapods*) has recently become popular with Darwinists because of the recent find of the *Tiktaalik* fossil (Daeschler et al. 2006). This fossil has been hailed as the "missing link that solves the mystery of evolution" (*The Guardian* 6 April 1986). It is the most newly found evolutionary intermediate between fish and amphibia (Rogers 2011, Dawkins 2010, Coyne 2009) and is claimed as further evidence of evolution (i.e., Common Descent).

Fossils of five extinct animals have been placed on the evolutionary line from the fish to the tetrapods. They are, in the presumed chronological order from oldest to youngest, *Eusthenopteron*, *Panderichthys*, *Tiktaalik*, *Acanthostega*, and *Ichthyostega*. In accordance with the neo-Darwinian paradigm, these fossils have been placed in an order that would illustrate a progression in the evolutionary transition from fish to tetrapods. The progression is of course predicated on the *assumption* of Common Descent, so it can hardly qualify as *evidence* of Common Descent. As in all examples of evolutionary transition, other interpretations are possible. Indeed, other interpretations may be more

---

6 The claim that the fossils are "transitional" is a Darwinist illusion. See below.

sensible, and I shall give an example. The Common-Descent interpretation is, in any case, unlikely to be true since random mutations and natural selection cannot account for it.

If one is bound to the neo-Darwinian paradigm, then one is compelled to interpret these fossils as a succession of animals gradually evolving from water-living fish toward land-living amphibia. But that is not the only interpretation of the fossil data, nor is it even a reasonable one. In light of the built-in capacity of animals to adapt by responding to environmental inputs, which we have seen in the previous chapters, I suggest that these animals may have been adapting independently to different environmental niches, perhaps in a way similar to the finches that diverged from a single parental stock.

The neo-Darwinian paradigm is quite misleading since its validity depends on the ability of the theory to account for the information buildup, and the theory cannot account for it. According to the dating of these fossils, these five species of animals have been determined to have lived at approximately the same time since the dates assigned to their fossils overlap. They have all been dated to the late Devonian period. The estimated dates span the 25-million-year interval from 385 million years ago (mya) to 360 mya. Fossil tracks of a tetrapod, which (according to their picture) should be *later* than all five of the above animals, have been found in Poland and dated to about 397 mya (Niedźwiedzki 2010), which is *earlier* than all of them. The dating strongly indicates that these species should be considered contemporaries. Note that the fossils that have been dated are only those that have been *found*. Future fossil finds might turn out to be dated before and after the dates given above. The above five fossils may well have been offshoots of the same species adapting to different environmental conditions through the NREH.

Consider the following hypothetical case. Suppose that finches become extinct and a future paleontologist comes across an island with fossil finches of various beak sizes. Suppose, furthermore, that he has a pet theory that early finches had short stout beaks and gradually evolved to birds with long thin beaks. He could collect fossils that showed the "early" beaks and some with the "later" beaks, as well as fossils in between. With the same logic as Darwinists have used on the fish-to-amphibian fossils, he could make a case for his pet theory. Of course he would be wrong in concluding that the long thin beaks evolved gradually from the short stout beaks. Just as his conclusion about finch evolution would be incorrect, so present-day speculation about the fish-to-amphibian evolution may be incorrect.

Robert Carroll (1995), a prominent paleontologist at McGill University, has noted that with regard to the supposed transition from aquatic to terrestrial animals:

> ... skeletal features that were once thought to be unique to land vertebrates appeared in a succession of aquatic, semiaquatic and semiterrestrial vertebrates, while features commonly attributed to fish were retained in animals classified as amphibians. ... The distribution of primitive and derived characters differs from lineage to lineage, showing that many features were evolved or lost *convergently*. (My emphasis)

What Carroll has said here is that some features once thought to belong only to land animals have been found in several of the fishes and some features once thought to belong only to fish were found in some land animals. To account for these fossils of aquatic and terrestrial animals, Darwinists invoke convergent evolution. I would suggest that the features inferred from these fossils, rather than being the result of convergent evolution, may well be the result of a built-in adaptive response to a changing environment, according to the NREH. These animals may not have descended one from the other at all, but they may well have all come from the same original stock and have been simultaneously adapting to various environmental niches just as the finches did. Both my suggestion and that of the Darwinists are, of course, speculations because we are dealing with events that happened in the past to which we have no experimental access.

In summary, here are the two explanations of the fossil record (mine and the Darwinists'):

- One invokes convergent evolution based on random mutations of the animals' DNA, a concept for which there is no observational evidence and no theoretical backing. (Darwinist interpretation)
- The other invokes the built-in capability of animals to adapt to a new environment, relying on phenomena that are observed today. (My suggestion)

Which would you say is the better explanation of these fossils? Although I cannot be sure of my interpretation, it at least shows that the Common-Descent picture does not have a monopoly on fossil-data interpretation; there are surely better interpretations. The fossil record is not a support for Common Descent.

Richard Dawkins (2010) likes to compare the interpretation of the circumstantial evidence of the fossil record with the way a detective interprets the evidence of a crime. If circumstantial evidence is to be relevant to a crime, however, there must be a theory of how the crime was committed — a theory that explains how the evidence is connected with the crime. The same applies to interpreting fossil evidence. Fossils alone cannot substantiate Common Descent. There must be a theory that describes how the evolution could have happened. But *there is no theory* — only stories. The fossils do not shed any light on how evolution can add information, and that of course is the issue that a theory of Common Descent is obligated to address.

Consider a hypothetical murder mystery where the victim was found dead on the floor of the library of his home with a gunshot wound in the head. A gun was found on the floor in the same room. The case is brought to court and the prosecuting attorney speaks.

"This is an open and shut case, your Honor. Exhibit A is the murder weapon as has been proved by ballistic tests comparing the bullet extracted from the victim with similar bullets fired from the same gun.

Furthermore, the gun is owned by the accused and has been shown to have his fingerprints on it. The circumstantial evidence is overwhelming. I recommend the jury find him guilty of murder."

The defense attorney then rises to speak.

"The circumstantial evidence alone is insufficient to warrant a guilty verdict, your Honor. To make the circumstantial evidence meaningful in this case, the prosecution must show how the accused could have carried out the murder. The accused is confined to a wheelchair and we have had expert testimony that he cannot get out of the chair by himself. May I remind the court that within less than a minute after the shot was heard, family members burst into the room to find the victim dead while the accused was in his wheelchair in his room one floor above. Unless the prosecution can give a plausible account of how the accused could have performed the murder and returned to his room a floor above within less than a minute, the circumstantial evidence cannot be used to convict him. I therefore urge the jury to find him innocent."

The same situation prevails with the attempt to use fossil evidence to support Common Descent. The fossils are circumstantial evidence requiring a theory that can account for how Common Descent could have occurred. Such a theory must account for how the information in living organisms could have been built up in the process of Common Descent. Neither the neo-Darwinian theory, nor any other theory presently known, is able to do this. One must therefore conclude that fossil evidence does not support Common Descent.

* * *

Geographical Distribution

The geographical distribution of plants and animals is another Darwinist argument for Common Descent based on theology and not on science. I have shown in the two preceding chapters, that animals and plants have a built-in capability of long-term adaptation to a new environment. A stimulus from a new environment can cause them to alter their genome to adapt to that environment — no random mutations are required. In Chapter 3, I have given examples of heritable changes in plants and animals induced by the environment. Living organisms are endowed with the ability to function efficiently in a variety of environments. Different versions of the same animal or plant can be found in various places on earth because each has undergone a different environmentally stimulated heritable change. Darwinists try to use these biogeographical data to argue for evolution by claiming that a Creator would not do it that way. Jerry Coyne wrote about plants:

> Why would a creator put plants that are fundamentally different, but look so similar, in diverse areas of the world that seem ecologically identical? Wouldn't it make more sense to put the same species of plants in areas with the same type of soil and climate? (Coyne 2009, 91)

And about animals, he wrote:

> If animals were specially created, why would the creator produce on different continents fundamentally different animals that nevertheless look and act so much alike? (Coyne 2009, 92)

He went on to write:

> The biogeographical evidence for evolution is now so powerful that I have never seen a creationist book article or lecture that has tried to refute it. (Coyne 2009, 88)

I hope he reads this book. I not only *try* to refute his argument, I demolish the argument for evolution from biogeography. Richard Dawkins wrote:

> Why would an all-powerful creator decide to plant his carefully crafted species on islands and continents in exactly the appropriate pattern to suggest, irresistibly, that they had evolved and dispersed from the site of their evolution? (Dawkins 2010, 270)

The "pattern" suggests nothing of the sort. It is only a Darwinist auto-suggestion.

What Coyne and Dawkins are saying here is that they "know" that random mutation and natural selection can produce different plants and animals with similar characteristics in different locations. The Creationist alternative, according to them, is for the Creator to fashion separate plants and animals in each and every location, and they consider that absurd.

Evidently, however, the Creator is cleverer than Coyne or Dawkins. He indeed seems to have "carefully crafted" information in His species giving them the ability to respond to environmental stimuli to alter their own genome to adapt to new environments. He then evidently let them wander where they will with an ability to adapt. It is presumptuous of Coyne, Dawkins, and their ilk to invoke amateurish theological arguments to support what they claim is a scientific hypothesis. Competent theologians have considered their questions centuries ago and have arrived at much different conclusions. Rabbi David Luria derived from the Talmud and the Midrash the necessity of animals to evolve. He wrote (Luria 1852, Ch. 23):

... באמת נ"ל שאף שהמינים רבים כמ"ש, הנה כללי סוגי המין הגבוה י"ל כדברי הפר"א שאינן אלא שס"ה, שהרבה מינים שאנו רואים לפנינו נפרדים ומשונים למיניהם, מעיקרן נבראו ונכנסו במין א', וכמ"ש בחולין (סג:) מאה עופות וכולם מין איה הן, ומעיקר בריאתן לא נברא כ"א המין והסוג הגבוה (שכללן אינו אלא שס"ה) ואח"כ לפי משכנם בארצותם ומאכלם ומקריהם נפרדו ונשתנו לעוד מינים רבים.

[... Indeed, it appears to me that even though the number of species is large, as we have explained, they are merely subgroups of higher categories that can be taken, according to Pirkei D'Rabbi Eliezer, to number only 365, because many species that we see today as separate and distinct were originally created as a single family. As stated in the Talmud Tractate Chullin (63b), "a hundred kinds of birds, all of which are of the family of the kite (*Accipitridae?*)," were not originally created. Rather, the higher category (of which there are only 365) was created and then split up and differentiated into many species according to their niche in their territory, and according to what they ate and what they experienced.] (My translation)

Rabbi Luria's theological treatment of the geographical distribution of living organisms agrees with the data of geographical distribution, which is not the case with the theology of Coyne or Dawkins. Coyne and Dawkins both seem to be unaware of the fact that animals and plants are able to change and adapt to new environments in response to an environmental input. The entire premise of their argument for evolution based on the geographical distribution of plants and animals is fallacious, and their argument is invalid.

Rabbi Luria describes a limited type of evolution, which is unlike Common Descent. There were 365 basic families of animals created, and the same number of bird families. All the others evolved from these, but not through random mutation and natural selection. Rather, they evolved by undergoing natural changes induced by their environment — new territory, new food, and other stimuli that may have acted on them. Lurian evolution is much like what I have described as the NREH in Chapter 2. His theological conclusion agrees with the data of the geographical distribution of animals. Once again, my purpose in bringing this theology here is not to support my arguments. Those arguments are well supported by biological data. I am bringing this theology only to contrast it with the theology that Coyne, Dawkins, and others use in their attempt to argue for evolution.

* * *

A Chain of Adaptive Mutations Not Supporting Common Descent

As I have noted, Common Descent, according to neo-Darwinian theory, requires long sequences of random mutations each of which must, on the average, add at least a small amount of new information to the gene pool that was not previously in the biosphere. If the Darwinian story were true, there would have to have been a huge number of information-adding random mutations to get the information and complexity now found in living organisms. These mutations would have had to have been built one on the other to achieve a sequence of increasing adaptivity.

If many such long sequences of mutations really do occur, then we should be able to observe some of them happening now. We should be able to find at least fragments of such sequences occurring both in the laboratory and in the wild even today. We should also be able to envisage how such fragments could be extended to indicate the workings of Common Descent. Finding such fragments would lend some support for Common Descent. If we had a fragment sequence of only a hundred of the many millions of adaptive mutations in the many different sequences, we could envisage it extended to the many millions that would be required for Common Descent. But we have not found a sequence of a hundred; we have not found a sequence of even ten, or even three, that have added information to the genome. The scientific literature records no example of even one random mutation that adds heritable information to the genome. That's damning evidence against Common Descent.

Jerry Coyne (2009) wrote about an experiment that shows a sequence of three mutations that look like they may be what I just said we haven't found. These three mutations successively add a new capability to a population of bacteria. Coyne described an experiment in which a gene was deleted from the bacterium *E. coli*. That gene encoded the enzyme that starts the breakdown of lactose, and without it the bacteria could not feed on lactose. The bacteria with the missing gene were then placed in an environment in which lactose was the only food source so they couldn't grow. After a while, however, a mutation appeared that permitted them to grow slowly. A while later a second mutation appeared that made the growth a little faster. Still later, a third mutation appeared that permitted still faster growth.

Superficially, this experiment looks like it shows some real evolution. One might think that these three mutations could be a fragment of the kind of long sequence that I said we have not seen. Coyne attributed this experiment to Barry Hall and his group at the University of Rochester. I couldn't check the details of this experiment because he cited no specific reference. His list of references for that particular chapter contains only one reference to Hall, which is a paper by him alone in 1982, which was long before Hall went to Rochester. In that paper, Hall does describe an experiment he did with *E. coli* from which a gene was deleted, but in that experiment a latent gene was activated that took over the job of the deleted gene. I discussed this experiment in my earlier book.

Curiously, Hall's 1982 paper does contain a description of a set of experiments done by others that was remarkably similar to the one Coyne described. These experiments were not on *E. coli* but on a soil bacterium. It just so happens that I described this set of experiments in my earlier book. I will not repeat that full discussion here, but I will summarize it. For more detail and references please see that discussion.

A population of soil bacteria that normally feed on the sugar ribitol was placed in an environment containing only the sugar xylitol, which they cannot use. They cannot feed on it for two reasons: First, they don't have an enzyme that can break down xylitol, and second, they don't have an enzyme that can transport the xylitol into the cell. So the bacteria did not grow at all on xylitol. After a while, however, a mutation occurred that allowed them to grow slowly on xylitol. Some time later, a second mutation occurred that allowed them to grow a little faster. Then a third mutation occurred that allowed still faster growth. But still, the growth was not as fast as the original population (the wild type) grows on ribitol.

The first of those three mutations was a point mutation that disabled the repressor of the RDH gene,[7] which encodes for the RDH enzyme, which breaks down ribitol. Disabling the repressor causes the RDH-enzyme synthesis to go at full blast, making large amounts of the enzyme. Although the RDH enzyme is specific to ribitol, it does happen to have some small activity on xylitol.[8] While normally this activity is too small to allow the bacteria to live on xylitol, the large numbers of this enzyme molecule, produced in the absence of the gene repressor, do permit small growth on xylitol. Still, since the cell has no

---

7 RDH = ribitol dehydrogenase
8 The structure of xylitol is very similar to that of ribitol.

transport enzyme for xylitol, only few molecules of xylitol can get into the cell by diffusion.

The third mutation,[9] also a point mutation, disabled the repressor of the transport gene encoding the transport enzyme of another sugar, D-arabitol. The normal role of this enzyme is to transport D-arabitol into the cell. It turns out that it happens to work well on xylitol but the gene is induced only by D-arabitol. The cell therefore does not make this enzyme if D-arabitol is not present. This third mutation disabled the repressor for this transport gene permitting it to make large quantities of the enzyme without having to be induced. Because of the large numbers of the transport enzyme, many xylitol molecules were taken into the cell, permitting faster growth.

The second mutation was not in a repressor, but it was a point mutation in the RDH gene itself. This mutation raised the activity of the RDH enzyme on xylitol and lowered it on ribitol. This may look like information is being gained, but closer examination shows that information is actually being lost. Testing this modified enzyme on another, related sugar showed that what this mutation has actually done was to lower the specificity of the enzyme (for more details see my previous book, *Not By Chance!*).

This loss of specificity is a loss of information — just like an address on an envelope "Washington, D.C." contains less information than the address "1600 Pennsylvania Ave., Washington, D.C." because the former is less specific than the latter. Lowering specificity is a loss of information just as raising the specificity requires adding information. Moreover, an enzyme must be specific to the reaction it catalyzes. The specificity of an enzyme is no less important to the cell — and perhaps even more important — than the level of its activity. An enzyme that will accept any molecule as its substrate can be harmful. For one thing, it wastes resources such as energy and nutrients that are not needed. For another, the unnecessary products it makes may interfere with

---

[9] You'll notice that I skipped the second mutation. I'll come back to that.

what the cell is supposed to be doing. For an enzyme to be useful to the cell, it must limit its activity to its proper substrate. As Bone, Silen, and Agard (1989) wrote, "One of the fundamental functions of an enzyme is to provide specificity by limiting the range of substrates which are catalytically productive." Losing the specificity of enzymes is not a formula for achieving Common Descent.

Thus we see that this chain of three mutations loses information with each mutation even though each helped the soil bacteria metabolize a new sugar that it had previously been unable to use. These mutations are not of the kind that can build Common Descent. In each case, the bacterium gained an advantage from each mutation, not because any information was added or because something new was created, but because something was disabled or degraded. The first and third mutations disabled a repressor protein, and the second mutation degraded the specificity of an enzyme. Sequences of mutations of disabling and degradation cannot create the new information required for Common Descent. A repressor is an important element in cellular control. It ensures that the gene it controls is OFF when its product is not needed. Disabling a repressor will let the gene be ON all the time. Although in the above case, when xylitol was the only nutrient, disabling the repressor helped the cell survive, under ordinary conditions the cell would not compete well with non-mutated cells. More mutations of this sort would just mean more disabling and more degradation, which is not a formula for Common Descent.

Thus, it turns out that not a single random mutation has been found that adds information to the genome.

* * *

Another Failed Darwinist Argument

Darwin suggested that living things strongly compete when their numbers rise to exceed their resources. This competition, according to him, leads to a natural selection of the most fit. Darwin got this idea from

Thomas Malthus (1798),[10] who was concerned about the human population eventually outstripping the food supply. Darwin applied this to animals and derived from it his theory of natural selection. If the Malthusian speculation were true about animals, we should see animals living under miserable conditions and always on the verge of starvation. Under conditions such as these, only the hardiest individuals would survive and procreate. But here Darwin was wrong.

Richard Dawkins tried to fashion this discussion into an argument for evolution. His argument (again, theological) was that a Creator would have planned the ecosystems "with the welfare of the whole community of wild animals at heart." Under the Darwinian paradigm, however, the individual animals would have evolved through natural selection to be concerned only with themselves and their progeny. Predators would tend to overhunt their prey, rather than be "prudent" predators the way humans try to preserve their resources for the future (although we are not entirely successful) by passing conservation laws and limiting hunting and fishing. He writes:

> "… shouldn't we expect wild predators, like wolves or lions, to be prudent predators too? The answer is no. No. No. No." (Dawkins 2010)

His argument is that this is what a Creator would be expected to do, but evolution, he claims, works otherwise.

Well, for once it appears that here Dawkins got his theology right, because that seems to be exactly what the Creator did! But this time Dawkins's biology is wrong. Predators do not generally overexploit their resources. In a restricted area such as an island, wolves limit their population size when the deer population goes down. In general, wolf and deer populations have an equilibrium toward which they tend (Messier and Crête 1985, Messier 1985, Messier 1994, Eberhardt and Peterson 1999, Eberhardt 2000, Messier and Joly 2000). The wolves do

---

10 In a letter to A. R. Wallace, Darwin wrote: "I came to the conclusion that selection was the principle of change from the study of domesticated productions; and then, reading Malthus, I saw at once how to apply this principle." (Darwin 1859)

not overexploit the deer population and drive it to extinction, but instead they limit their own population to ensure that the deer continue to serve as food for future wolf generations (Wynne Edwards 1965 & 1986, Bergerud 1983). They live, so to speak, only on the income from their resources and preserve the principal for posterity.

If animals behaved the way Dawkins tells it, they would be living a miserable existence. But the fact is that they usually live well.

Wynne-Edwards (1986) has suggested that animal populations are generally kept in check not by extrinsic forces such as mass starvation or disease, but by intrinsic controls built into the animals themselves. This phenomenon may be surprising and even amazing to most people, but biologists studying animals in the wild have reported this kind of control operating in a variety of populations.

Plants also do not proliferate in a field to the point where they become overcrowded. They do not engage in a "struggle for existence" for natural selection to preserve those that pass the survival test and destroy those that don't. Plants tend to control their populations by sensing the density of the planting. When the growth is dense, plants produce less seeds; when growth is thin, they produce more seeds (Bradshaw 1965). Both plants and animals have built-in programs for avoiding overexploitation of their resources — something we humans must do by legislation.

Here Richard Dawkins fell into a trap of his own making. His argument for evolution is, in reality, an argument against it.

* * *

Evolution Happening Now

Industrial Melanism

Darwinists offer examples of evolution-in-action implying that these support Common Descent. These examples include the famous one of the evolution of melanism in peppered moths (*Biston betularia*). Before the industrial revolution in Great Britain and before factories began to burn coal and smoke up the air on a large scale, the peppered moth was light in color and dark ones were rare. The moths would generally rest on lichen-covered tree trunks where they were camouflaged against the light background. As the atmospheric smoke turned the lichens dark, the dark (melanic) moths became more numerous until they were about 90% of the moth population. This transformation was interpreted as having been the result of natural selection, and it may well have been. Against the dark background of the soot-covered tree trunks, the dark moths would be much less visible than the light ones to the birds that preyed on them.

In the 1960s England began to clean up its air by requiring the use of cleaner-burning fuel, and the lichens began to lose their dark covering. As the lichens regained their original light color, the percentage of dark moths began to decrease and that of the light moths began to increase until the dark moths have become rare once more. Kettlewell (1955, 1956) performed experiments that reinforced the presumption that the rise of the dark moths, and their subsequent decline, was the result of natural selection.

Darwinists bring this example as evidence of evolution, with the implication that it supports Common Descent. The evolution of "industrial melanism," as it is called, however, can be dismissed as giving no support for Common Descent at all because there was no new mutation. The necessary variant forms were already in the population (Bishop and Cook 1975), so no new information has been added to the biosphere. There is no way a phenomenon like the evolution of melanism could lead to Common Descent, no matter how many millions of years it acts.

Antibiotic resistance
The evolution of antibiotic resistance has been for some time the Darwinists' favorite example for "demonstrating" evolution (Common Descent). Superficially their case looks good. Antibiotics date only from about 1930 with the discovery of penicillin (Fleming 1929), followed by the development of a method to produce it with high yield (Chain

et al. 1940). Antibiotics were first introduced to the public in 1942 to cure bacterial infection (Levy 1992, 4), and by the mid 1940s the first strains appeared of *Staphylococcus* resistant to penicillin (Fisher 1994, 15). Just a few years after antibiotics were introduced, resistant strains of the pathogens were found to have already evolved. As each new antibiotic was discovered and put into use against pathogenic bacteria, resistant strains soon followed. The argument then goes, with a wave of the hand, like this: If a small but significant evolutionary change like antibiotic resistance can evolve in only a few years, then surely in a million years huge evolutionary changes must occur. Darwinists expect this argument to support Common Descent.

An examination of the phenomenon of antibiotic resistance, however, shows it lends no support at all to Common Descent (Spetner 1997, 138-143). Antibiotics are natural molecules produced by some microorganisms for the purpose of killing other hostile microorganisms. A microorganism that makes an antibiotic must, itself, be resistant to the antibiotic it makes. For this purpose it is typically endowed with a battery of genes that code for a resistance mechanism. Most useful antibiotics have come from soil bacteria (D'Costa et al. 2006). How bacteria have acquired this resistance initially is not known, nor can neo-Darwinian theory shed any light on it. Antibiotic resistance genes have been found to predate the use of antibiotics by at least many thousands of years (D'Costa et al. 2011). Moreover, bacteria are known to be able to transfer genetic material to other bacteria through HGT (see above). On occasion, copies of the genes for resistance can find their way from a type of bacterium that is normally resistant to a type that is not normally resistant. When that happens, the recipient bacterium becomes resistant. This is indeed evolution, but it is a limited evolution of the population-change type. It is not the Common-Descent type of evolution.

The resistance genes already exist in the biosphere. No new information has appeared in the biosphere through this type of evolution of antibiotic resistance. Common-Descent evolution cannot be achieved by this procedure even if it were repeated innumerable times in succession, because no new information would be built up. This method of evolving antibiotic resistance therefore lends no support for Common Descent.

Sometimes, however, antibiotic resistance can indeed appear through a random mutation — a DNA copying error, which would bring something new to the biosphere. This kind of change looks like it might satisfy the requirements for Common Descent, so I shall give a brief description of it here, although I have already dealt with it in my previous book.

As an example, let us look at how a bacterium acquires resistance to streptomycin through a random mutation. All cells, whether of bacteria or of plants or animals, contain organelles called ribosomes, whose function it is to make protein according to instructions from the DNA of a gene. Proteins are large molecules, consisting of long chains of small molecules called amino acids, and are essential to all living things. They function as enzymes, which catalyze all the chemical reactions in a cell — each chemical reaction catalyzed by a specific enzyme. Proteins can also serve as structural elements. Often, and maybe even always, a structural protein functions also as an enzyme. For an enzyme to perform its function, it must have a specific sequence of amino acids.

A ribosome is an organelle within a cell that manufactures protein. It makes a protein by putting together a chain of amino acids according to the instructions in the DNA. A segment of the DNA is transcribed into an RNA molecule that matches the DNA nucleotide by nucleotide. This RNA is called messenger RNA because it carries the DNA message to the ribosome. The ribosome translates the message in the DNA into amino acids according to the genetic code. Three nucleotides translate into one amino acid. Accordingly, the ribosome constructs a chain of amino acids to form a protein.

The antibiotic streptomycin, for example, acts on a bacterial cell by attaching to a ribosome at a site to which it matches, the way a key fits into a lock. When the streptomycin molecule attaches to this site, it interferes with the ribosome function and causes it to make mistakes leading to incorrect, dysfunctional or nonfunctional, protein. The errors it causes prevent the cell from growing, reproducing, and eventually from living. The important feature of streptomycin, and indeed of all other antibiotics, is that it kills bacteria but does not harm the mammalian host. Streptomycin kills the bacterial cells that are infecting you without killing your own cells. It discriminates between the cells of the bacteria and the cells of the host by its specific attachment to a matching site on the bacterial ribosome, a site not found on the host's ribosomes.

A bacterium will gain resistance to streptomycin if a point mutation occurs in the gene coding for the protein in the ribosome, ruining the matching site, destroying the specificity of the protein, and preventing a streptomycin molecule from attaching. If the streptomycin cannot attach to the matching site, the bacterium is resistant. Just one mutation in the portion of the DNA coding for the matching site can mess up the site so the streptomycin cannot attach. It turns out that any one of several mutations in that portion of the DNA will grant the bacterium resistance (Gartner and Orias 1966). Note that this type of resistance is caused by a single random point mutation, but it cannot serve as an example of mutations that can support Common Descent. One cannot expect mutations destroying specificity, no matter how many of them there are, to build information and lead to Common Descent. Destruction of specificity does not add information — it destroys it. One cannot add information by destroying it, no matter how many times one repeats the process. I have previously (Spetner 1997) compared trying to build up information in this manner to the merchant who was losing a little money on each sale but thought he could make it up on volume. The acquisition of antibiotic resistance is indeed evolution, but only a limited form of it. It cannot lead to Common Descent.

No example of antibiotic resistance in bacteria adds information to the biosphere. To become resistant, the bacteria either pick up ready-made resistance genes from other bacteria or they undergo a mutation that destroys information. Antibiotic resistance cannot therefore be an evolutionary example that could support Common Descent because a chain of such mutations, no matter how long, does not add information and thus cannot lead to Common Descent. The Darwinists' favorite example of evolution fails to pass muster.

* * *

Vestigial Organs, Junk DNA, and Pseudogenes

It has been customary for Darwinists to use any biological feature whose function is not known as a "proof" of Common Descent. Organs whose functions are not known are declared to be vestiges of ancestral forms that once had a use but no longer have one. Darwinists would point to an organ in humans, in particular, that had no apparent purpose and would declare it was vestigial and must have been a leftover from an ancestor in whom the organ was functional. Coyne (2009, p. 56) wrote, "'vestigial organs' … make sense only as remnants of traits that were once useful in an ancestor." His "proof" is the naïve theological argument, "a Creator would not put useless organs into His creatures." Theological arguments, however, are not science.

Moreover, the argument is flawed because one cannot possibly tell with any assurance that an organ has no use. We may well have not yet found its use. An organ that now seems to have no function may in the future be found to have a function. Looking back on the history of the designation of vestigial organs, we see this to have indeed happened time and again. Functions have continually been found for organs that were once thought to be vestigial. About 120 years ago, Robert Wiedersheim published a list of 86 human organs he called vestigial (Wiedersheim 1895). Gradually, one by one, these were later found to be functional, exposing the logical fallacy of the vestigial-organ argument. According to Darwinists there are still a few left, but the trend is clear — the vestigial-organ list is gradually being depleted.

Jerry Coyne (2009, 60 ff.) thinks the vermiform appendix in humans must still be called vestigial. In fact he considers it "the most famous" of the vestigial organs. It turns out, however, that the appendix is not only functional but it plays a critical role. It is a storage place for gut bacteria that repopulate the gut after diarrheal illness. Pathogenic bacteria, in small numbers, cannot usually compete with the normal benign resident gut bacteria, and they therefore do not usually pose a danger to us. But when the gut bacteria are depleted during a diarrheal illness, any pathogens present would have free reign and would multiply rapidly causing serious illness. The benign gut bacteria stored in the appendix prevent this catastrophe (Bollinger et al. 2007, Laurin et al. 2011). Coyne condescends to acknowledge the appendix "may have some small use. ... as a refuge for useful bacteria," but he minimizes its importance and implies it is, through evolution, on the way out. This is at odds with the latest research showing that the appendix plays a critical role. "Your appendix could save your life," writes Rob Dunn (2012). Coyne is holding on to the appendix with his fingernails, so to speak, in the hope he won't lose one of the last vestigial organs with which to "prove" evolution. It appears, however, that it has nevertheless slipped from his grasp. Using apparently functionless organs as proof for evolution is indeed a losing game.

As vestigial organs are moving offstage, the so-called "junk DNA," which I mentioned in Chapter 1, has taken over the role of "vestigial" DNA to "prove" evolution. The most outstanding of junk DNA for Darwinists are what are called *pseudogenes*. A pseudogene is a section of DNA that looks like a gene but doesn't code for protein or RNA because it apparently cannot be activated. Like vestigial organs, pseudogenes did not seem to have a function and were therefore hailed by Darwinists to be vestiges of genes that were once operational in an evolutionary ancestor but have since lost their ability to function. Jerry Coyne (2009, 66-67) says that from evolutionary theory,

> ... we can make a prediction. We expect to find, in the genomes of many species, silenced, or "dead" genes: genes that once were useful but are no

longer intact or expressed. In other words, there should be vestigial genes.

Funny thing. Coyne didn't make that prediction *before* pseudogenes were found. Darwinists like Coyne always seem to make their "predictions" after the fact. Instead of a *prediction*, it should be called a *postdiction*. Coyne continues with his naïve theology:

> In contrast, the idea that all species were created from scratch predicts that no such genes would exist.

Dawkins also writes about pseudogenes. He also says they are utterly useless and contribute nothing to the organism. According to him, they are good for only one thing (Dawkins 2009, 332-333):

> What pseudogenes are useful for is embarrassing creationists. It stretches even their creative ingenuity to make up a convincing reason why an intelligent designer should have created a pseudogene — a gene that does absolutely nothing and gives every appearance of being a superannuated version of a gene that used to do something ...

Judging from the history of vestigial organs, you would think a smart scientist would suspect pseudogenes might have a function. The latest research actually bears this out. Wen et al. (2012) have reported, "Pseudogenes are not pseudo anymore." In their conclusion, they write, "We believe that more and more functional pseudogenes will be discovered as novel biological technologies are developed in the future."

The living organisms we see around us are exceedingly complex and contain a vast amount of information. Analysis of the human genome, which has more than three billion nucleotides, has shown it has only about 20,000 genes (consisting of perhaps twenty million nucleotides). Could all the complexity of the human body be produced by only 20,000 genes? Perhaps it could if those genes were controlled to function in several different ways. To do that, the genes would have to be under sophisticated control, and that control would have to be carried out by some of the DNA in the genome.

Until very recently, the conventional wisdom among biologists was that more than 99% of the genome was "junk" because it did not code for protein. This "junk" was said to be "left over" from early ancestors and just carried along as the DNA was replicated. This "junk" was used as an argument for Common Descent with the argument that a Creator would not put junk into the genome.

But much of this "junk" has recently been shown to be active in the cell. Even though it does not code for protein, it turns out that much of it does code for RNA, which seems to play a regulatory role in the cell.

Project ENCODE was established by the National Human Genome Research Institute in 2003 to discover the function, if any, of all DNA in the genome. The project was started with a pilot study to examine 1% of the genome. The results of this study indicated that more than half the genome had some biochemical function (ENCODE Project Consortium 2007). The next stage of the project was to examine the entire genome. These results were published in 2012 in which the junk-DNA myth was debunked with overwhelming data (ENCODE Project Consortium 2012). They raised the minimum fraction of functional DNA from half up to 80%. A spokesman for the project, Joseph R. Ecker (2012), wrote:

> ... 80% of the genome contains elements linked to biochemical functions, dispatching the widely held view that the human genome is mostly "junk DNA." The authors report that the space between genes is filled with enhancers (regulatory DNA elements), promoters (the sites at which DNA's transcription into RNA is initiated) and numerous previously overlooked regions that encode RNA transcripts that are not translated into proteins but might have regulatory roles.

Another spokesman, Inês Barroso (2012), wrote:

> Results from the ENCODE project show that most of these stretches of DNA harbour regions that bind proteins and RNA molecules, bringing these into positions from which they cooperate with each other to regulate the function and level of expression of protein-coding genes.

Some Darwinists, who were upset with the possibility of losing their "junk DNA," rebutted this report with the quibble about how the consortium defined the word "functional" (see, e.g., http://www.genomicron.evolverzone.com/2012/09/a-slightly-different-response-to-todays-encode-hype/ and http://pandasthumb.org/archives/2012/09/encode-hype-fro.html). They suggested the consortium's use of the word "functional" was not what should be meant by that word. They granted this 80% may indeed be chemically active in the cells, but they questioned if they were really of benefit to the organism. Whereas the consortium spokesmen suggested the products of the 80% might have regulatory roles in the cells, the above Darwinists countered merely by suggesting they might not.

W. Ford Doolittle (2013) also employed the definition-of-functional quibble to attack the conclusions reached by the spokesmen for Project ENCODE. But he added an additional argument. He argued that there has to be "junk DNA" because it is only with the concept of "junk DNA" that he and others feel they have resolved the C-value paradox. Without junk DNA they would no longer have their resolution of the paradox. In Chapter 5 I shall describe the C-value paradox and indicate how the NREH might help to resolve it without "junk DNA."

But I don't understand why those Darwinists have to be so upset about the ENCODE conclusions. Richard Dawkins showed them how to wriggle out of it in a public debate he had with the Chief Rabbi of Great Britain, Jonathan Sacks (Sacks and Dawkins 2012). When Rabbi Sacks pointed out that ENCODE results show that junk DNA is no more, Dawkins countered with:

> "I know there are some creationists who have jumped on it because they think it is awkward for Darwinism. Quite the contrary, of course. It is exactly what a Darwinist would hope for — to find usefulness in the living world."

Of course, in his book, Dawkins (2009) had written quite the opposite:

> "… it is a remarkable fact that the greater part (95 percent in the case of humans) of the genome might as well not be there, for all the difference it makes."

The "junk" DNA and pseudogenes are going the way of the vestigial organs. They cannot be called upon to support Common Descent. Arbitrarily labeling them useless does not argue for Common Descent.

* * *

The Argument for Evolution From "Poor Design"

Darwinists use what they call "poor design" in living systems as an argument for Common Descent. The argument is theological and says that a Creator would not be expected to adopt a poor design. They say that it would be expected for evolution, on the other hand, to come up with designs that are suboptimal.

The so-called "inverted" retina of the vertebrate eye is their favorite example. The other two "poor designs" often cited as evidence of evolution are the convoluted path of the laryngeal nerve in mammals and the convoluted path of the vas deferens. The design of these organs appear to a naïve observer (and when it comes to the fascinating world of living organisms, even the greatest scientist is a naïve observer) to be flawed and unworthy of an omniscient Creator. Coyne (2009) uses the laryngeal nerve as an argument for evolution, Rogers (2010) uses the vas deferens, and Dawkins (2011) uses all three.

The retina of the eye is the screen on which the eye's optical image is focused. Nerves (bundled in the optic nerve) convey the image information to the brain. One would think the nerves (neurons) should come off the back of the retina, the side opposite the one having the image. But in vertebrates they, surprisingly, come off the front where the image is formed. A naïve observer would think this to be a poor arrangement because the neurons might interfere with the light falling on the retina. The Darwinists say, with their usual theological argument, that if it were designed by an omniscient Creator, it would surely have the nerve connections coming out of the back side of the retina. Since they come off the front side, against one's expectation, Darwinists conclude there is no omniscient Creator and therefore evolution must be true.[11]

---

11 Of course, they have no precise idea of how evolution could have led to the

The laryngeal nerve supplies two-way communication from the brain to the larynx (voice box). Although one might think this nerve should take the shortest path, it does not. It takes a convoluted path in mammals — branching off from the vagus nerve, it goes down through the neck, ducks under the aorta, and loops back up to the larynx, traveling what seems to be an unnecessary extra distance. This long nerve gives off many branches that innervate tissues along the way. In the giraffe, the laryngeal nerve's extra distance is extreme — about 15 feet. This, too, seems to be a poor design, and the Darwinists claim only an evolutionary process of random mutations and natural selection could account for it.[12] An omniscient Creator, they say, would surely have done a better job. The same argument is used for the vas deferens, which carries sperm from the testes to the penis and also takes what appears to be an unnecessarily circuitous route.

With regard to the inverted retina, it has recently been discovered that, rather than being a dumb design, it is actually remarkably clever. The cleverness is not in the neurons on the image side of the retina, but in the glial cells, which always accompany neurons. The neurons are transparent and do not interfere with the passage of light, but the glial cells aid the process of vision by channeling the light. The glial cells of the retina are long and thin and propagate light as in an optical fiber (Franze et al. 2007), and have been called "ingeniously designed light collectors." Amichai Labin and Ezra Ribak of the physics department of the Technion (Israel Institute of Technology) have shown by simulation and calculation that the glial cells improve the optical resolution of the retina and compensate for chromatic aberration (Labin and Ribak 2010). Had the optic neurons come off the back side of the retina, these advantages would not have accrued.

With regard to the laryngeal nerve and the vas deferens, I frankly

---

development of the retina. Typical of their vague arguments, they don't know what mutations would be necessary to generate the nerve network, or if it could be done at all through a sequence of adaptive mutations.

12 Here, too, Darwinists cannot explain how evolution could do this.

do not know why they take those circuitous routes. But it seems to me presumptuous to declare the designs dumb simply because we do not understand what advantages they might have. We do not know what further research may show.

Let me give an example from engineering. The feedback parameters of a feedback-control system determine how rapidly the system will respond. The more stable the system, the more sluggish is its response. If the feedback parameters are adjusted to raise the speed of response too high, the system will go into oscillation and become unstable.

The F-16 fighter aircraft, for example, was designed and built in the late 1960s and early 1970s. It was purposely designed to be slightly unstable to increase its rate of response and make it more maneuverable. The aircraft is prevented from going into oscillation by having a flight-control computer between the pilot's hand-and-foot controls and the aerodynamic control surfaces themselves. This computer automatically trims the controls to maintain stability, whereas without it the pilot would have to do the trimming manually. If he didn't do it just right, the aircraft could go out of control.

Now suppose an engineer of the 1950s, who did not understand the role of the flight-control computer, were to examine the control-system. He would conclude that the design was dumb because the aircraft was unstable. He would say that he would have designed it differently. He would of course be mistaken. The design is actually a very smart one because it permits high maneuverability, using the computer to maintain stability.

It is presumptuous of the Darwinists to suppose they understand the design of biological organs well enough to pass judgment on the quality of the design. So the Darwinist argument from poor design is rather an argument from incomprehensibility. The Darwinist looks at the path of the laryngeal nerve and says, "I cannot comprehend why it takes this circuitous route. It seems like a poor design. I would have designed it differently. So would an omniscient Creator, if there is One. Therefore there isn't One, and these organs are the products of evolution with its random mutations." The likelihood is that he is wrong and we may yet discover the benefit of these designs, just as we have of the inverted retina.

* * *

I have shown in the earlier chapters that there is no proper theory of evolution in the sense of Common Descent. In this chapter I have shown that what is offered as the best evidence for Common Descent does in fact not support it at all. The lack of a proper theory and the lack of proper evidence for Common Descent must lead us to conclude that *evolution, in the sense of Common Descent, is simply not true.*

Chapter 5

Epilogue

THE RELEVANCE OF EVOLUTION HAS BEEN EXAGGERATED FAR beyond what could be supported by theory and data. I have shown in this book that evolution is limited by the evolutionary capabilities built into organisms. Science has so far not shed any light on where those capabilities came from. The Darwinian theory of random variation and natural selection is unable to account for them.

From all I have said here, one must conclude that the claim of Common Descent, and consequently macroevolution, is not supported by evidence and is therefore not believable. Yet the Darwinists have been calling it a *fact* (Gould 1981, Lewontin 1981, Futuyma 1986, 15). I have shown it has no theoretical backing and there are no data that can be said to support it. Add to this that it is counterintuitive, and it is impossible to understand how it can be called a "fact."

There is no argument, however, about evolution in the sense of population change. That kind of evolution occurs and is observable. I have critiqued only the evolution of Common Descent, the emptiness of which I have exposed in this book. There is no scientific backing for supposing humans evolved from a type of primitive ape, nor is there any scientific backing for supposing life evolved from a primitive cell. To the layman, "evolution" means Common Descent. Therefore, in popular parlance, *evolution* has hereby been dispatched. Contrary to what leading scientists have written, and contrary to what is written in publications of the National Academy of Sciences, Science has not shown life is just an accident and Science has not shown that we have evolved in a purely natural way from some primitive cell.

Francisco Ayala has drawn a parallel between the Copernican and the Darwinian revolutions (Ayala 2007). The Copernican Revolution was the shifting of conventional wisdom from holding that the earth was the center of the universe to holding that the sun was the center of the solar system, comprising the sun, the earth, and the planets. The Copernican Revolution began with the publication in 1543 of Nicolaus Copernicus's work, *De revolutionibus orbium celestium*, continued through the work of Kepler and Galileo, and reached its peak with Sir Isaac Newton's publication of his *Principia*. Newton established mathematical equations governing the motion of physical bodies, and in particular how they move under the influence of gravity. With these equations he was able to account for how everyday objects fall to the ground as well as to account for the motion of celestial bodies including the earth, the moon, and the planets. Ayala noted that these equations enabled Newton to bring the motions of physical objects, including the heavens, into the realm of science. Yet, he noted, until the publication of Darwin's *Origin of Species*, the existence and structure of living things were still accounted for only by supernatural causes.

Ayala called the disparity between these two types of explanation "conceptual schizophrenia," and credited Darwin with resolving the schizophrenia by bringing the study of living things into the realm of science. But there is a vast difference between the way Newton brought the study of celestial motion into science and the way Darwin is said to have brought the study of life into science. Newton established equations from which the motion of celestial bodies could be precisely calculated and compared with observation, whereas Darwin proposed only a vague process of chance through imaginary scenarios. Newton's theory has been rigorously checked against observation. Darwin's "theory" has not been checked in anything like so rigorous a manner. Ayala was guilty of a gross exaggeration when he wrote, "Darwin completed the Copernican Revolution by drawing out for biology the notion of nature as a lawful system of matter in motion that human reason can explain without recourse to supernatural agencies." Darwin did no such thing.

Charles Darwin's novel concept of natural selection was a good idea, and has been shown to work in some cases. But in trying to extend it all the way back to early life and concocting Common Descent from it, he overreached. Natural selection has indeed been shown to work in the laboratory and perhaps in some cases confirmed by observations in the field, but it cannot account for Common Descent. Most of the evolution we observe in nature is driven by the organisms' built-in response to environmental inputs, according to the NREH as described in Chapter 2. Natural selection creates nothing; it can only reject. The creative part of evolution in neo-Darwinian theory is relegated to random mutation. This is somewhat analogous to the research department in a manufacturing company. The research scientists invent and management chooses which inventions to pursue. The creativity and the gaining of information in the genome is entirely in the hands of mutation (the research department). Common Descent requires the information in all living organisms to have been built up by natural means. But it has never been shown that this can be done by random variation and natural selection, and there is good reason to believe it cannot.

Unless there is a mutation in an individual that can lead to an adaptive improvement, natural selection has nothing to work on. The mutation may have happened now or at any time in the past, but it must be in the population for natural selection to work. Neo-Darwinian theory says that these mutations are random. That means their occurrence cannot be described by equations as Newton had done with mechanics; they can be described only in terms of probabilities. In place of solving equations, one must instead calculate the probability that there is an adaptive mutation or recombination available at each stage in evolution. Darwinists tacitly assume such mutations are always available, but there is no basis for that assumption. They look at a population change in the past, possibly as inferred from the fossil record, or inferred from molecular analysis, and say it was the result of random-mutation/natural-selection that happened to be adaptive to an environmental change. They always assume the right random mutations are available at the right time. We have seen in Chapter 3 that if such population changes occurred, they were more reasonably the result of the organisms' built-in response to environmental inputs. The latter are known to occur. Random nucleotide copying errors adding information to the genome is something that has never been observed. It is merely a vain hope of the Darwinists.

Darwinists are wont to obscure the element of randomness in evolution. They continually say natural selection makes the process nonrandom. That is only a half-truth. Natural selection can give the process a direction, but according to their theory it can work only on the mutations that they insist are random. Because of the inherent randomness embodied in the mutations, their theory is obligated to address the issue of the chance of occurrence of the right mutations. Are the right mutations very likely, as Darwinists tacitly assume, or are they highly improbable, as some of the analyses of mine and others have indicated? Darwinists have not addressed this issue. Without showing these mutations to be very probable, there is no theory of Common Descent, otherwise popularly known as *the theory of evolution*.

The claim that Common Descent is well established is what Darwinists like to call the scientific basis of the proposition that the origin of life was a purely natural phenomenon, sometimes called *naturalism*. It forms the basis of the evangelical atheistic agenda of Richard Dawkins, Jerry Coyne, and their ilk. It is, today, the theoretical basis of atheism. Richard Dawkins (1986) wrote that Darwin made it possible to be an intellectually fulfilled atheist. On what a pitifully weak basis does his atheism rest!

For centuries, conventional wisdom of the Western world held the universe was created. Those who were uncomfortable with a Creator could take some refuge in believing the universe was infinitely old[1]

---

[1] That belief can no longer be reasonably held today in light of the evidence for the

and it was never created and therefore didn't need a Creator. But they had a problem with life. Life is not infinite. There are no living beings of infinite age so life had to start somewhere. Small living things like maggots could apparently generate spontaneously (this was before Pasteur), but these were in any case supposed to emerge from decaying organic matter so that couldn't solve the problem of the origin of life from inert matter. The existence of animals and plants indeed seemed to need a Creator. Conventional wisdom of the time held "... living things belonged outside the realm of material principles ... If life obeyed any laws, they were supernal and not bound to the physics of inert substance." (Simpson 1960) Creation held sway even with would-be atheists mainly because a Creator was necessary to explain the existence of life.

The publication of Darwin's *Origin of Species* brought a dramatic change to this philosophical position. Although for some time there had been speculations about a natural origin of life, Darwin's work offered what seemed to be a scientific account of how life could have evolved naturally from some simple beginning. As George Gaylord Simpson (1960) has pointed out, although parts of Darwin's theory are now known to have been wrong, the theory was adequate at the time in the sense of being convincing. The essential point, according to Simpson, was that Darwin convinced the scientific community that material causes of evolution are possible and can be investigated scientifically. They have been investigating now for more than 150 years but have come up empty-handed. That's okay; investigating is what Science is supposed to do. But what is not okay is that they are convinced they have found that Common Descent is explainable within the science that we know, while all they really have is a half-baked

---

Big Bang in the Hubble red shift and the cosmic background radiation. "Cosmologists thought they had a workaround. Over the years, they have tried on several different models of the universe that dodge the need for a beginning while still requiring a big bang. But recent research has shot them full of holes. It now seems certain that the universe did have a beginning." (*New Scientist*, January 13, 2012)

theory with no good supporting data, as we have seen in Chapter 4.

The concept of living things evolving was in the intellectual air well before Darwin published the *Origin*, but it had not yet been generally accepted by the scientific community. It was Darwin's argument that made it acceptable. Darwin's argument was essential for the acceptance of evolution. Simpson wrote (1960):

> It does seem to me highly improbable that the fact of evolution would have been accepted so widely and quickly if it had been unaccompanied by an explanatory theory.

John Maynard Smith (1993) wrote:

> ... the fact of evolution was not generally accepted until a theory had been put forward to suggest how evolution had occurred, and in particular how organisms could become adapted to their environment; in the absence of such a theory, adaptation suggested design, and so implied a creator. It was this need which Darwin's theory of natural selection satisfied. He was able to show adaptation to the environment was a necessary consequence of processes known to be going on in nature.

Darwin's theory of variation and natural selection indicated continuity among living organisms. It implied that one form of organism could flow almost seamlessly into another. The assumed continuity encouraged the hope that perhaps this continuity could extend back to non-living matter as well. If that were so, then life could have had a natural origin and the need for a Creator could be successfully eliminated.

In the late 1860s, Thomas Huxley and his colleagues, who had formed what they called the X Club of London, believed that life could form spontaneously from nonliving matter, but they were reluctant to publicize their speculations because they did not want to be identified with advocates of spontaneous generation, which Louis Pasteur had recently discredited with his famous experiment (Strick 1999). They thought protoplasm, the contents of a cell, was a homogeneous mass that was the basis of living organisms. Huxley called it "the physical basis of life." The chemical composition of protoplasm was unknown.

Vitalists held protoplasm contained some (mysterious) "vital" component that made it alive, while others held its chemical composition alone is what made it alive. Huxley and his colleagues were seeking to demonstrate that protoplasm could be formed from inert chemicals with the hope they could close the conceptual gap between inert chemicals and living organisms.

Darwin's *Origin*, as well, held out the promise that we could find a way to account for the origin of life without the need for a Creator, but this promise has not been fulfilled. More than half a century ago, George Gaylord Simpson (1960) wrote:

> At a recent meeting in Chicago, a highly distinguished international panel of experts was polled. All considered the experimental production of life in the laboratory imminent ...

So much for the predictions of experts! Today's experts are still saying the same thing, but we are no closer to producing life from inert chemicals than we were in 1960.

The publication of the *Origin* was the watershed event that led to encouraging the hope of proving that life could have originated in a purely natural way and that there was no need for a Creator. Although there were atheists before the publication of the *Origin* who were looking for ways of eliminating the need for a Creator, most scientists during that time believed in a Creator, as did the general population of the Western world.

The argument in *The Origin of Species* is not a theory in the modern sense because it was too vague on the details of how evolution could happen. The argument essentially was that evolution seemed plausible, but the apparent plausibility was a result of the limited knowledge of biology that prevailed at the time. I would guess if Darwin's arguments would have waited until now to be presented they would not have had the acceptance they received in 1859. Today (in fields other than evolution) we are more demanding about what constitutes a scientific theory. We would have at least demanded a probability calculation to show the random events that are crucial to the theory to have a high enough probability to make the theory reasonable. But what happened is that Darwin's theory (*speculation* would be a more appropriate term) was accepted in the nineteenth century and has been subjected to little serious cross-examination since. Almost every new discovery in biology has been uncritically hailed as additional "proof" of evolution. But after the critical examination of Common Descent presented in this book, we see that these so-called "proofs" of evolution do not provide any support for Common Descent.

We therefore return to Square One, pre-Darwin: We have no natural explanation for the existence of life. Those who would opt for Creation have their explanation, but it is grounded in faith and not in science. Those who refuse to opt for Creation can choose to believe that future advances in evolutionary theory may someday vindicate their position of a natural origin of life. But this belief is also grounded in faith, not science.

One cannot honestly say evolution in the sense of Common Descent is a scientific theory, despite the Darwinists' hyperbolic statements about evolution — the kind of statements no scientist would think of making in another field. Richard Dawkins wrote about evolution (Dawkins 2009):

- It is the stunningly simple but elegant explanation of our very existence and the existence of every living creature on the planet.
- Darwin's idea is arguably the most powerful ever to occur to a human mind.
- Any theory of life has to explain how the laws of physics can give rise to a complex flying machine like a bird or a bat or a pterosaur, a complex swimming machine like a tarpon or a dolphin, a complex burrowing machine like a mole, a complex climbing machine like a monkey, or a complex thinking machine like a person. Darwin explained all of this with one brilliantly simple idea — natural selection, driving gradual evolution over immensities of geological time.
- … if you meet somebody who claims not to believe in evolution, that person is ignorant, stupid or insane …

All this is, of course, nonsense in light of what I have shown in this book. With dogmatic certainty, Simpson (1960) wrote:

> ... man's ancestors were apes or monkeys .... It is pusillanimous if not dishonest for an informed investigator to say otherwise.

Professor Vincent Cassone, chair of the University of Kentucky's biology department, is reported to have told the *Herald-Leader* (Cassone 2012):

> There is more evidence for evolution than there is for the theory of gravity, than the idea that things are made up of atoms, or Einstein's theory of relativity. It is the finest scientific theory ever devised.

The finest scientific theory ever devised?! Hardly. Why are these hyperbolic statements made about evolution and not about any other science? I suspect it is because Darwinists are insecure in their beliefs about evolution and are whistling in the dark, trying to reassure themselves as well as the public that their ideas are not foolish.

What should we be teaching about evolution in the science classes of our public schools? We must certainly stop teaching that Common Descent is a fact. Darwinists have been saying "Evolution is not just a theory. It's a fact!" for too long. It is neither a fact nor is it even a proper scientific theory. Rather than their contention "Evolution is not just a theory," I would invert the order of two of those words and say: "Evolution is *just not* a theory!" Intimidation must stop. Students should not be cowed into believing that all educated people must hold life is just the product of purely natural causes and of random events. There is no scientific basis for such statements.

For too many decades the public educational system in the Western world has been advocating an atheistic *Weltanschauung,* falsely bestowing on it the imprimatur of Science. I have not seen a serious study of the matter so I hesitate to suggest what this does to young minds, but I doubt it has been beneficial. Teaching falsehoods cannot be beneficial. It should be universally agreed that false concepts not be taught. We must stop teaching the false concept of Common Descent as if it were a fact. Public and private schools should not be intimidated into teaching Common Descent. Free schools in Britain were recently reported to be threatened to have their funding cut if they do not teach Common Descent (Burns 2012).

Should we teach evolution in science classes? Yes, we should teach that populations evolve. Populations do indeed evolve. But we should stop teaching Common Descent and that population change results from random mutations and natural selection. We should stop teaching that life is a natural phenomenon because we really do not know that. We should stop intimidating students to believe that humans are descended from some primitive ape-like beast because we don't know that, either.

We should teach that plants and animals evolve to become better adapted to their environment, but teach how rapidly that has been seen to occur. Teach that organisms have a built-in capacity for adapting to a changing environment. Teach about the built-in capacity of organisms for self genetic engineering (e.g., Shapiro 2011). Examples like Darwin's finches should be taught not as an indication of a broad evolutionary principle of Common Descent, but as an expression of built-in responses to environmental stimuli. The many examples of evolution that happen today, which are the only solid examples of evolution we have, should be taught without suggesting they imply Common Descent, because they don't.

Shall we teach Creation in the public schools? Definitely not! The subject is best handled in the home or within a religious environment. Creation is a religious, not a scientific, subject. Unless Creation were to become verifiable within science, it belongs in the realm of religion. In any case, the evolution I suggest we teach poses no argument with Creation, but different religions may have different takes on Creation and teaching it would run the danger of segueing into teaching religion itself.

The approach I advocate to teaching evolution forgoes the concept of Common Descent together with attempts to develop a universal phylogenetic tree that will encompass all living things. In their place it offers insights leading to studying the only kind of evolution we observe — evolution in action today. The nonrandom evolutionary hypothesis (NREH) described in Chapter 2 is that environmental inputs can induce heritable adaptive changes in living organisms, and this hypothesis matches observations.

Further Research

It seems to me that if evolution studies were to drop Common Descent and adopt the NREH as its model instead, exciting new avenues of research would open up. The barren concept of Common Descent would be replaced by a much more fruitful one. The NREH suggests many new lines of research. The skeleton of the hypothesis, which is beckoning to be filled out, is that environmental inputs stimulate the release of first messengers in the organism, which in turn trigger second messengers within the cells. These in turn cause genetic rearrangements that can lead to adaptations in the phenotype. Some of the exciting questions that follow are:

- How do environmental inputs cause the release of first messengers? What are the molecular details of this process?
- Stress, stemming from the environment, is known to cause genetic rearrangements. What kinds of stresses have this effect?
- By what molecular action do these stresses stimulate the first messengers, or hormones?
- Do different kinds of stress trigger different hormones? If there are different ones, how many are there and how does this number and their types depend on the species?
- For a given species, what is the range of environmental inputs that produce a phenotypic change?

- How often are these changes adaptive to the environment? How often are these changes heritable?
- What kinds of genetic rearrangements can appear as the result of environmental inputs?
- For any given species, what are its capabilities of responding to environmental inputs?
- If we apply an environmental stimulus, which portion of the genome becomes active?
- Other genetic engineering.

In answering these questions, many new ones would likely suggest themselves and they could lead to many fruitful research projects. As is usual with research, attempts to answer questions can lead to new insights and many further questions.

Resolving the C-value Paradox

Research on the NREH could lead to a resolution of what is called the C-value paradox. The mass of DNA in a haploid cell (a gamete)[2] of an organism is called the organism's "C-value."[3] The C-value is generally expressed in units of a picogram (pg), which is a trillionth of a gram. This is a convenient unit because it leads to relatively small numbers for most C-values. There are 978 million DNA base pairs in one picogram. The C-value of humans is about 3.5 pg, which corresponds to about 3.4 billion base pairs.

The DNA in the cells of an organism is generally understood to contain information necessary for its development and function. It therefore seems reasonable to suppose that the amount of an organism's DNA is some kind of a measure of its complexity. Yet there does

---

2 The cells of sexually reproducing organisms have their chromosomes in pairs (except for the gametes). The members of each pair come one from each parent. The C-value of these organisms is the mass of only one set of its chromosomes, which is half of the total DNA.

3 Don't ask what the "C" stands for; Hewson Swift (1951), who introduced that parameter, did not say why he used "C" and it doesn't really matter. The term has caught on, and that's all that matters.

not seem to be a consistent relationship between the C-value and the apparent complexity of the organism. This inconsistency has been considered paradoxical. It seemed particularly troubling to find the C-values of some animals were larger — even much larger — than the C-value of humans. The C-value of the blue-spotted salamander (*Ambystoma laterale*), for example, is 81 pg, or 23 times as large as that of the human. The C-value of the gulf-coast waterdog (*Necturus lewisi*) is 120 pg, or 34 times that of the human. There is no reason to suppose the development and functioning of these animals are more complex than those of the human. It therefore seems strange that these animals should have so much more DNA than does the human. This puzzle is what is known as the C-value paradox (Thomas 1971).

Equally puzzling is that even some closely related species of animals or plants *have* very different C-values. For example, take the two toads, Couch's spadefoot toad (*Scaphiopus couchii*) and the European fire-bellied toad (*Bombina bombina*). The first has a C-value of only about 1.0, while the second has a C-value of more than 12. Similarly, of the two frogs, the ornate burrowing frog (*Limnodynastes ornatus*) and the ornate horned frog (*Ceratophrys ornata*), the first has a C-value of about 1.0 while the second has a C-value of more than 13. There are many other frogs, toads, and other animals among which there are pairs of species with similar morphology that have widely different C-values. The spread in C-value among the entire class of amphibia is more than a factor of 120.[4]

Why should animals that are very similar have such different C-values? If one of them can develop and function with the smaller DNA content, why does the other need so much more? And not just a little more, but many times more?

The consensual answer lies in the presumed whimsical result of random mutation and natural selection. The answer is anchored in the

---

4 These C-values were taken from the Animal Genome Database at www.genomesize.com.

concept of "junk" DNA (Eddy 2012, Gregory 2005). Conventional wisdom is that the disparity of C-values can be ascribed to large amounts of useless, or "junk," DNA in some species, and less in others. Some species just seem to accumulate more junk than others.

But this answer fails in light of the results of the ENCODE project, which I described in Chapter 4, and which reported that the "junk" DNA is not really junk. I also noted there that the Darwinists are upset with this conclusion. Some rebut the report with the argument that the "junk" is needed to resolve the C-value paradox (Eddy 2012, Doolittle 2013). The argument is circular: the existence of "junk" DNA is said to resolve the C-value paradox, while *the C-value paradox is used as an argument for the validity of* "junk" DNA. A circular argument is not a solution.

It seems to me, however, that the NREH, as described in Chapter 2, offers the possibility of properly solving the C-value paradox. The built-in capability of responding to an environmental input requires information, and this information would most reasonably reside in the genome. I shall call the DNA that carries this information the N-DNA, where the "N" stands for the NREH. The N-DNA would normally not be active. If one were to delete any portion of it, one would normally not see any effect. In Chapter 2 I cited an experiment in which an ultraconserved section of DNA was deleted from a mouse with no discernable effect (Ahituv et al. 2007). Because this section is ultraconserved it must have an important function. But since that function was not revealed by its deletion, it must be designated for something other than normal living or development. It is reasonable to suppose it belongs to the N-DNA. The effect would become apparent only in the presence of an appropriate environmental stimulus, which would occur only in a changing environment.

I am suggesting that the N-DNA is the extra DNA in species with high C-value. Species that must adapt to a wide variety of environmental conditions should have a large amount of N-DNA. Those that adapt to only a narrow variety of conditions would have less N-DNA. Here is an interesting research project suggested by the NREH.

Endangered Species

The NREH can shed light on the important subject of rescuing endangered species. In the United States there are more than a thousand species of animals and plants officially listed as endangered. Conservation organizations claim there may be more than ten times that number. Endangered species are those that are in danger of becoming extinct. One reason a species can become extinct is that its habitat may deteriorate to where the species can no longer cope. During global cooling, for example, unless the animals move to a warmer region they may become extinct. Other species that can survive under the new colder conditions may take their place. The species that take their place may be entirely new species that moved into the newly available niche, or they may be modifications of the former species. The modifications are likely to be the result of the species's response to the new environmental inputs, as postulated by the NREH. They are far more likely to result from the NREH than from random mutations. In such a case there is no real loss of a species because the new species is just the old species modified by its built-in ability to alter its genome in response to the environmental stress. Under the right conditions, this change might even be reversed.

When a habitat deteriorates, conservationists sometimes construct a new habitat and populate it with animals (or plants) rescued from the deteriorated environment. For example, a subpopulation of the endangered Appalachian elktoe mussel has been transferred to a North Carolina Wildlife Resources Commission hatchery, with the hope that these animals can flourish sufficiently to remove them from the endangered list (National Park Service Website 2012). Another example is the Devil's Hole pupfish, mentioned in Chapter 2, that were transferred to a special pupfish refuge (Lema 2008). The National Wildlife Refuge System has 59 refuges, which are home to 280 endangered species (National Fish and Wildlife Service 2012).

An apparent problem with the refuges is that they may not exactly match the original habitat. If a refuge does not match the original habitat, the species may change (evolve) through its built-in capability to adapt. But since it is still basically the same species as before, a species has not really been lost. The NREH thus gives us a more optimistic view of the endangered-species problem.

Such a view might also enable us to better manage our financial resources. Saving endangered species does not come free. It costs big money. Conservationists have estimated that the cost of rescuing and maintaining all endangered species would be more than $75 billion annually (McCarthy et al. 2012). This is not petty cash. It behooves us to do our endangered-species saving in the most efficient manner. To make the cost affordable, we should try to focus on the major species, where "major" means a single representative population from which a group of species can evolve, as postulated by the NREH. We must expect, though, that in the act of rescuing a population, the species may in any case be altered.

*　*　*

Evolution is much more limited in its scope than has been generally assumed. All the real evolution we observe, as I have noted, is not explained by random mutations and natural selection — rather, it is obscured by it. Let us drop Darwin's "great" idea, which has really been quite barren. W. R. Thompson, a fellow of the Royal Society of London, and who was the Director of the Commonwealth Institute of Biological Control, Ottawa, wrote in 1963 in his introduction to the Everyman's Library edition of *The Origin of Species*:

> I do not contest the fact that the advent of the evolutionary idea, due mainly to the *Origin*, very greatly stimulated biological research. But it appears to me that owing precisely to the nature of the stimulus, a great deal of this work was directed into unprofitable channels or devoted to the pursuit of will-o'-the-wisps.

It would be much more fruitful to pursue an understanding of how evolution really happens. We should study how living organisms respond to environmental stimuli rather than just to assume it happens through DNA copying errors. The dogmatic assumption of Common Descent stifles profitable research on evolution. Let's study evolution as it really happens and not take the cop-out of assuming the processes are so long-term that one cannot observe them happening. It would be exciting to discover just how environmental stimuli lead to changes in the genome. More and more, we see new examples of rapid adaptation of a population to a new environment. In Chapter 3, I have given many examples where several species rapidly achieve the same adaptation. These examples are labeled "convergence" and are called "surprising" (Princeton Univ. 2012), "spectacular" (Liu et al. 2010), "remarkable" (Bossuyt and Milinkovich 2000, Thorne et al. 2004), and "striking" (Neuweiler 2003, Fleischer et al. 2008). They are "surprising," but only under the neo-Darwinian paradigm. Under the NREH, they are not surprising but expected.

Let us stop pretending evolutionary events occur through random mutations and find out how they really occur. Biology has had an exciting ride in the twentieth century. Biology in the twenty-first century portends to be even more exciting.

Acknowledgments

My thanks to several friends who have read the manuscript and gave me useful comments. These include Laszlo Bencze, who read through the early manuscript and made many valuable stylistic suggestions. I thank James Shapiro for reading an early manuscript and making several critical and useful comments. My thanks also go to Sanford Sampson, Shalom Kellman, Betsy Siewert, Cornelius Hunter, and John Calvert, who read an early draft and made helpful suggestions. Special thanks to my son Abba, who read and edited every version of the manuscript. I also thank my granddaughter Rivka Schonfeld for drawing the figure on page 55.

References

Abzhanov, A., M. Protas, B. R. Grant, P. R. Grant, and C. J. Tabin (2004) *Bmp4* and Morphological Variation of Beaks in Darwin's Finches. *Science* **305**(5689): 1462–1465.

Ahituv, Nadav, et al. (2007) Deletion of ultraconserved elements yields viable mice *PLoS Biology* **5**(9): 1906–1911.

Andersson, Jan O. (2009) Gene transfer and diversification of microbial eukaryotes. *Annual Review of Microbiology* **63**: 177–193.

Anfinsen, C. B. (1959) *The Molecular Basis of Evolution* New York: Wiley.

Ayala, Francisco J. (2007) Darwin's greatest discovery: Design without designer *Proceedings National Academy of Science USA* **104 suppl. 1**: 8567–8573.

Baldwin, J. Mark (1896) A New Factor in Evolution. *The American Naturalist* **30** (355): 441–451, 536–553.

Barluenga, Marta, Kai N. Stölting, Walter Salzburger, Moritz Muschick and Axel Meyer (2006) Sympatric speciation in Nicaraguan crater lake cichlid fish. *Nature* **439**: 719–723.

Barton, N. H., J. S. Jones, and J. Mallet (1988) No barriers to speciation. *Nature* **336**: 13–14.

Bastian, H.C. (1870) Protoplasm *Nature* **1**: 424–426.

Bejerano, Gill, et al. (2004) Ultraconserved Elements in the Human Genome. *Science* **304**: 1321–1326.

Bergerud A. T. (1983) Prey switching in a simple ecosystem. *Scientific American* **249** December.

Bishop, J. A. and L. M. Cook (1975) Moths, melanism and clean air. *Scientific American* **232** January.

Blount, Zachary D., Christina Z. Borland, and Richard E. Lenski (2008) Historical contingency and the evolution of a key innovation in an experimental population of *Escherichia coli*. *Proceedings National Academy of Science USA* **105** (23): 7899–7906.

Bollinger, R. R., A. S. Barbas, E. L. Bush, S. S. Lin, and W. Parker (2007) Biofilms in the large bowel suggest an apparent function of the human vermiform appendix. *Journal of Theoretical Biology* **249**: 826–831.

Bossuyt, Franky and Michel C. Milinkovich (2000) Convergent adaptive radiations in Madagascan and Asian ranid frogs reveal covariation between larval and adult traits. *Proceedings of the National Academy of Sciences USA* **97**(12): 6585–6590.

Bradshaw, A. D. (1965) Evolutionary significance of phenotypic plasticity in plants. *Advances in Genetics* **13**: 115–155.

Brown, J. K. and A. Kodric-Brown (1979) Convergence, competition and mimicry in a temperate community of hummingbird-pollinated flowers. *Ecology* **60**: 1022–1035.

Brown, W. L. Jr. and E. O. Wilson (1956) Character Displacement. *Systematic Zoology* **5**: 49–64.

Buck, L. and R. Axel (1991) A novel multigene family may encode odorant receptors: a molecular basis for odor recognition. *Cell* **65**: 175–187.

Burns, J. (2012) Teaching evolution key to free school funding deal. BBC News http://www.bbc.co.uk/news/education-20547195

Canfield, Donald E., Alexander N. Glazer, and Paul G. Falkowski (2010) The Evolution and Future of Earth's Nitrogen Cycle. *Science* **330**: 192–196.

Caporale, L. H. (2003) Darwin in the Genome. New York: McGraw-Hill.

Carroll, R. L. (1995) Between fish and amphibian. *Nature* **373**: 389–390.

Carroll, S. P., H. Dingle and S. P. Klassen (1997) Genetic differentiation of fitness-associated traits among rapidly evolving populations of the soapberry bug. *Evolution* **51**: 1182–1188.

Carroll, S. P., A. P. Hendry, D. N. Reznick and C. W. Fox (2007) Evolution on ecological time-scales. *Functional Ecology* **21**: 387–393.

Case, Ted J. (1997) Natural selection out on a limb. *Nature* **387**: 15–16.

Cassone, V. (2012) Reported on the website of the National Center for Science Education http://ncse.com/news/2012/08/kentucky-legislators-assailing-evolution-007511

Chain, E. et al. (1940) Penicillin as a Chemotherapeutic Agent. *Lancet* **239**: 226–228.

Chamary, J. V. and Laurence D. Hurst (2009) The Price of Silent Mutations. *Scientific American* **300**: 46–53.

Chen, Yiming, Robin Lowenfeld and Christopher A. Cullis (2009) An environmentally induced adaptive (?) insertion. *International Journal of Genetics and Molecular Biology* **1 (3)**: 038–047.

Cody, M. L., and J. M. Overton (1996) Short-term evolution of reduced dispersal in island plant populations. *Journal of Ecology* **84(1)**: 53–61.

Cohen, P. (1988) Review Lecture: Protein Phosphorylation and Hormone Action *Proceedings The Royal Society (London)* **B234 (1275)**: 115–144.

Conant, S. (1988) Saving endangered species by translocation. *BioScience* **38**: 254–257.

Coyne, J. A. (2009) *Why Evolution Is True.* Penguin Group USA.

Crick, Francis (1988) *What Mad Pursuit* New York: Basic Books

Cullis, Christopher A. (2005) Mechanisms and Control of Rapid Genomic Changes in Flax. *Annals of Botany* **95**: 201–206.

D'Costa, Vanessa M., Katherine M. McGrann, Donald W. Hughes, and Gerard D. Wright. (2006) Sampling the antibiotic resistome. *Science* **311**: 374–377.

D'Costa, Vanessa M. et. al. (2011) Antibiotic Resistance is Ancient. *Nature* **477**: 457–461.

Dávalos, Liliana M., Andrea L. Cirranello, Jonathan H. Geisler, and Nancy B. Simmons (2012) Understanding phylogenetic incongruence: lessons from phyllostomid bats. *Biological Reviews* **87(4)**: 991–1024.

Daeschler, E. B., Shubin, N. H. and Jenkins, F. A. Jr (2006) A Devonian tetrapod-like fish and the evolution of the tetrapod body plan. *Nature* **440**: 757–763.

Darwin, C. (1859) Letter from Charles Darwin to A. R. Wallace, in Darwin, Francis and A. C. Seward, eds. (1903) *More Letters of Charles Darwin*, 2 vols. London: John Murray, vol. 1, p. 118, as cited by Evans (1984).

Darwin C. (1860) Letter from Charles Darwin to Asa Gray, February 1860, in Darwin F., ed. (1888) *Life and Letters of Charles Darwin*, 3 vols, London: John Murray, vol. 2, p. 273.

Darwin, C. (1871) Letter from Charles Darwin to Joseph Dalton Hooker on February 1st, 1871.

Dawkins, R. (1986) *The Blind Watchmaker*. New York and London: W.W. Norton.

Dawkins, Richard (2003) *A Devil's Chaplain*. New York: Houghton Mifflin.

Dawkins, R. (2009) *The Greatest Show on Earth: The evidence for evolution*. New York: Free Press.

Degnan J. H. and N. A. Rosenberg (2006) Discordance of species trees with their most likely gene trees. *PLoS Genetics* **2(5)**: 762–768.

Degnan JH and Rosenberg NA. (2009) Gene tree discordance, phylogenetic inference and the multispecies coalescent. *Trends in Ecology and Evolution* **24(6)**: 332–340.

Dermitzakis E. T., Reymond A. and Antonarakis S. E. (2005) Conserved non-genic sequences — An unexpected feature of mammalian genomes. *Nature Reviews Genetics* **6**: 151–157.

Diamond, J. M., (1996) Daisy gives an evolutionary answer. *Nature* **380**: 103–104.

Dickinson M (2008) Animal locomotion: a new spin on bat flight. *Current Biology* **18**: R468–R470.

Dieckmann, Ulf and Michael Doebeli (1999) On the origin of species by sympatric speciation. *Nature* **400 (6742)**: 354–357.

Dobler, S., S. Dalla, V. Wagschal and A. A. Agrawal (2012) Community-wide convergent evolution in insect adaptation to toxic cardenolides by substitutions in the Na,K-ATPase. *Proceedings of the National Academy of Sciences USA* **109(32)**: 13040–13045.

Dobzhansky, T. (1937) *Genetics and the origin of species*. New York: Columbia University Press.

Dobzhansky, T. (1938) The Raw Materials of Evolution. *The Scientific Monthly* **46(5)**: 445–449.

Doebeli, M. (1996) A quantitative genetic competition model for sympatric speciation. *Journal of Evolutionary Biology* **9**: 893–909.

Doolittle, W. F. (2013) Is junk DNA bunk? A critique of ENCODE. *Proceedings of the National Academy of Sciences USA* **110**: 5294–5300.

Drake, J. W. (1999) The distribution of rates of spontaneous mutation over viruses, prokaryotes, and eukaryotes. In Molecular Strategies in Biological Evolution, L. H. Caporale ed., *Annals of the New York Academy of Sciences* **870**: 100–107.

Duerden, J. E., (1920) Inheritance of callosities in the ostrich. *The American Naturalist* **54**: 289–312.

Dunn, Emmett Reid (1943) Lower Categories in Herpetology. *Annals of the New York Academy of Sciences* **44**: 123–131.

Eberhardt, L. L. and R. O. Peterson (1999) Predicting the wolf-prey equilibrium point. *Canadian Journal of Zoology* **77(3)**: 494–498.

Eberhardt, L. L. (2000) Reply: Predator-prey ratio dependence and regulation of moose populations. *Canadian Journal of Zoology* **78(3)**: 511–513.

Ecker, Joseph R. (2012) ENCODE explained: Serving up a genome feast. *Nature* **489**: 52–53.

Eddy, Sean R. (2012) The C-value paradox, junk DNA and ENCODE. *Current Biology* **22(21)**: R898–R899.

Eden, M., (1967) Inadequacies of the neoDarwinian evolution as a scientific theory, in Moorehead and Kaplan (1967), pp. 512.

Eldredge, N., and S. J. Gould (1972) Punctuated equilibria: an alternative to phyletic gradualism. In T. J. M. Schopf, ed., *Models in Paleobiology*. San Francisco: Freeman, Cooper and Company, pp. 82–115.

Emera, Deen and Günter P. Wagner (2012) Transformation of a transposon into a derived prolactin promoter with function during human pregnancy. *Proceedings of the National Academy of Sciences USA* **109(28)**: 11246–11251.

ENCODE Project Consortium (2007) Identification and analysis of functional elements in 1% of the human genome by the ENCODE pilot project. *Nature* **447**.

ENCODE Project Consortium. (2012) An integrated encyclopedia of DNA elements in the human genome. *Nature* **489(7414)**: 57 DOI: 10.1038/nature11247.

Erwin, D. H. (2004) One Very Long Argument. *Biology and Philosophy* **19**: 17–28.

Feder, J. L., C. A. Chilcote, and G. L. Bush (1988) Genetic differentiation between sympatric host races of the apple maggot fly *Rhagoletis pomonella*. *Nature* **336**: 61–64.

Fenchel, T., (1975) Character displacement and coexistence in mud snails. *Oecologia* **20**: 19–32.

Filchak, K.E. et al. (2000) Natural selection and sympatric divergence in the apple maggot *Rhagoletis pomonella*. *Nature* **407**: 739–742.

Fisher, Jeffrey A. (1994) *The Plague Makers: How we are creating catastrophic new epidemics — and what we must do to avert them*. New York: Simon & Schuster.

Fleischer R. C., H. F. James and S. L. Olson (2008) Convergent evolution of Hawaiian and Australo-Pacific honeyeaters from distant songbird ancestors. *Current Biology* **18**: 1927–1931.

Fleischer R. C., H. F. James and S. L. Olson (2008) Convergent evolution of Hawaiian and Australo-Pacific honeyeaters from distant songbird ancestors. *Current Biology* **18**: 1927–1931.

Fleming, A. (1929) On the antibacterial action of cultures of a Penicillium, with special reference to their use in the isolation of *B. influenzae*. *British Journal of Experimental Pathology* **10**: 226–238.

Franze et al. (2007) Müller cells are living optical fibers in the vertebrate retina *Proceedings of the National Academy of Sciences USA* **104(20)**: 8287–8292.

Fry, B. G., K. Roelants and J. A. Norman (2009) Tentacles of Venom: Toxic Protein Convergence in the Kingdom Animalia. *Journal of Molecular Evolution* **68**: 311–321.

Futuyma, Douglas J. (1986) *Evolutionary Biology*, 2nd ed., Sunderland MA: Sinauer Associates.

Gartner, T. K. and E. Orias, (1966) Effects of mutations to streptomycin resistance on the rate of translation of mutant genetic information. *Journal of Bacteriology* **91**: 1021–1028.

Giovannoni, S. J., Hayakawa, D. H., Tripp, H. J., Stingl, U., Givan, S. A., Cho, J.-C., Oh, H.-M., Kitner, J. B., Vergin, K. L. and Rappé, M. S. (2008) The small genome of an abundant coastal ocean methylotroph. *Environmental Microbiology* **10**: 1771–1782.

Gordon, Swanne P., David N. Reznick, Michael T. Kinnison, Michael J. Bryant, Dylan J. Weese, Katja Räsänen, Nathan P. Millar, and Andrew P. Hendry (2009) Adaptive Changes in Life History and Survival following a New Guppy Introduction. *The American Naturalist* **174(1)**: 34–45.

Gould, Stephen J. (1981) Evolution as Fact and Theory. *Discover* (May)

Gould, S. J. (1989) *Wonderful Life: The Burgess Shale and the Nature of History* New York: Norton.

Grant, P. R. (1986) *Ecology and Evolution of Darwin's Finches*. Princeton: Princeton University Press.

Grant, Peter R. and B. Rosemary Grant (2006) Evolution of Character Displacement in Darwin's Finches. *Science* **313**: 224–226.

Gregory T. R. (2005) The C-value enigma in plants and animals: a review of parallels and an appeal for partnership. *Annals of Botany* **95**:133–46.

Hall, B. G. (2004) Predicting the evolution of antibiotic resistance genes. *Nature Reviews. Microbiology* **2(5)**: 430–435.

Hall, Barry G. (1999) Transposable elements as activators of cryptic genes in *E. coli*. *Genetica* **107**: 181–187.

Hall, B.G. and P.W. Betts (1987) Cryptic genes for cellobiose utilization in natural isolates of *Escherichia coli*. *Genetics* **115**: 431–439.

Hall, B. G., S. Yokoyama, and D. H. Calhoun (1983) Role of cryptic genes in microbial evolution. *Molecular Biology and Evolution* **1**: 109–124.

Hazen Robert M. and James Trefil (2009) *Science Matters: Achieving Scientific Literacy* (rev. ed.) New York: Anchor Books.

Heled, Joseph and Alexei J. Drummond (2010) Bayesian Inference of Species Trees from Multilocus Data. *Molecular Biology and Evolution* **27(3)**: 570–580.

Hersh, Megan N., Rebecca G. Ponder, P.J. Hastings and Susan M. Rosenberg (2004) Adaptive mutation and amplification in *Escherichia coli*: two pathways of genome adaptation under stress. *Research in Microbiology* **155**: 352–359.

Hillenmeyer, M. E. et al. (2008) The chemical genomic portrait of yeast uncovering a phenotype for all genes. *Science* **320**: 362–365.

Himmelfarb, G. (1962) *Darwin and the Darwinian Revolution*. Garden City: Doubleday

Ho, M. W. and P. T. Saunders. (1979) Beyond Neo-Darwinism: An epigenetic approach to evolution. *Journal of Theoretical Biology* **78**: 573–591.

Hölldobler B. (2004) Ernst Mayr: the doyen of twentieth century evolutionary biology. *Naturwissenschaften* **91(6)**: 249–254.

Holliday, Robin (2006) Epigenetics: A Historical Overview (Review). *Epigenetics* **1(2)**: 76–80.

Hull, David L. (1970) Contemporary Systematic Philosophies. *Annual Review of Ecology and Systematics* **1**: 19–54.

Huxley, T.H. (1869) The Physical Basis of Life. *Fortnightly Review* (February). As cited by Strick (1999).

International Human Genome Consortium (2001) Initial sequencing and analysis of the human genome. *Nature* **409**: 860–921.

Iwama, G. K. (1998) Stress in Fish. *Annals of the New York Academy of Sciences* **851**: 304–310.

Jablonka, Eva (2009) Transgenerational epigenetic inheritance: Prevalence, mechanisms, and implications for the Study of heredity and evolution. *The Quarterly Review Of Biology* **84(2)**: 131–176.

Jackman, T., J.B. Losos, A. Larson and K. de Queiroz. (1997) Phylogenetic studies of convergent adaptive radiations in Caribbean Anolis lizards. pp. 535–557 in T. Givnish and K. Systma, Eds., *Molecular Evolution and Adaptive Radiation*. Cambridge Univ. Press.

Jacob, F. and J. Monod (1961) Genetic regulatory mechanisms in the synthesis of proteins. *Journal of Molecular Biology* **3**: 318–356.

Jacob, François (1977) Evolution and Tinkering. *Science* **196(4295)**: 1161–1166.

Jeffroy, Olivier, Henner Brinkmann, Frédéric Delsuc and Hervé Philippe (2006) Phylogenomics: the beginning of incongruence? *Trends in Genetics* **22(4)**: 225–231.

Johnson, Kevin P, Scott M Shreve and Vincent S Smith (2012) Repeated adaptive divergence of microhabitat specialization in avian feather lice. *BMC Biology* **10 (1)**: 52. DOI:10.1186/1741-7007-10-52.

Johnston, T. D. and G. Gottlieb, (1990) Neophenogenesis: A developmnental theory of phenotypic evolution. *Journal of Theoretical Biology* **147**: 471–495.

Kakudo, Shinji, Seiji Negoro, Itaru Urabe, and Hirosuke Okada (1993) Nylon Oligomer Degradation Gene, nylC, on Plasmid pOAD2 from a Flavobacterium Strain Encodes Endo-Type 6-Aminohexanoate Oligomer Hydrolase: Purification and Characterization of the nylC Gene Product. *Applied and Environmental Microbiology* **59(11)**: 3978–3980.

Kanagawa, K., M. Oishi, S. Negoro, I. Urabe and H. Okada (1993) Characterization of the 6-aminohexanoate-dimer hydrolase from *Pseudomonas sp.* NK87. *Journal of General Microbiology* **139**: 787–795.

Kanagawa, Kazuo, Seiji Negoro, Nobuo Takada, and Hirosuke Okada (1989) Plasmid dependence of *Pseudomonas sp.* Strain NK87 Enzymes that degrade 6-aminohexanoate-cyclic dimer. *Journal Of Bacteriology* **171**: 3181–3186.

Keating, Joseph C. Jr., (2002) The Meanings of Innate. *Journal of the Canadian Chiropractic Association* **46(1)**: 4–10.

Keeling, P. J., and J. D. Palmer (2008) Horizontal gene transfer in eukaryotic evolution. *Nature Reviews Genetics* **9**: 605–618.

Kettlewell, H. B. D. (1955) Selection experiments on industrial melanism in the Lepidoptera. *Heredity* **9**: 323–342. doi:10.1038/hdy.1955.36

Kettlewell, H. B. D. (1956) A resume of investigations on the evolution of melanism in the Lepidoptera. *Proceedings of the Royal Society* **B145**: 297–303.

Kinoshita, S., S. Kageyama, K. Iba, Y. Yamada, and H. Okada. (1975) Utilization of a cyclic dimer and linear oligomers of e-aminocaproic acid by *Achromobacter guttatus* KI72. *Agricultural and Biological Chemistry* **39**: 1219–1223.

Kinoshita, S., S, Negoro, M. Muramatsu, V. S. Bisaria, S. Sawada and H. Okada (1977) 6-Aminohexanoic acid cyclic dimer hydrolase. A new cyclic amide hydrolase produced by *Achromobacter guttatus* KI74. *European Journal of Biochemistry* **80**: 489–495.

Kinoshita, S., T. Terada, T. Taniguchi, Y. Takene, S. Masuda, N. Matsunaga, and H. Okada (1981) Purification and characterization of 6-aminohexanoic-acid-oligomer hydrolase of *flavobacterium sp.* K172. *European Journal of Biochemistry* **116**: 547–551.

Labin, A. M. and E. N. Ribak (2010) Retinal glial cells enhance human vision acuity. *Physical Review Letters* **104**: 158102. doi:10.1103/PhysRevLett.104.158102

Land, M. F. and R. D. Fernald (1992) The evolution of eyes. *Annual Review of Neuroscience* **15**: 1–29.

Laurin, Michel, Mary Lou Everett and William Parker (2011) The Cecal Appendix: One More Immune Component With a Function Disturbed By Post-Industrial Culture. *The Anatomical Record* **294(4)**: 567–579.

Lee, C. E. (2002) Evolutionary genetics of invasive species. *Trends in Ecology and Evolution* **17**: 386–391.

Lema, Sean C. (2008) The phenotypic plasticity of Death Valley's pupfish. *American Scientist* **96**: 28–36.

Levy, Stuart B. (1992) *The Antibiotic paradox: How Miracle Drugs are Destroying the Miracle*. New York: Plenum Press.

Levy, Asaf, Schraga Schwartz and Gil Ast (2010) Large-scale discovery of insertion hotspots and preferential integration sites of human transposed elements. *Nucleic Acids Research* **38(5)**: 1515–1530.

Lewontin, R. C. (1981) Evolution/creation debate: a time for truth. *BioScience* **31**: 559.

Liu, Yang, James A. Cotton, Bin Shen, Xiuqun Han, Stephen J. Rossiter, and Shuyi Zhang (2010) Convergent sequence evolution between echolocating bats and dolphins. *Current Biology* **20**: 1834–1839.

Losos, Jonathan B. (1990) A phylogenetic analysis of character displacement in Caribbean Anolis lizards. *Evolution* **44(3)**: 558–569.

Losos, Jonathan B. and Dolph Schluter (2000) Analysis of an evolutionary species-area relationship. *Nature* **408**: 847–850.

Losos, J. B. (2001) Evolution: A lizard's tale. *Scientific American* **284(3)**: 64–69.

Losos, J. B., K. I. Warheit, and T. W. Schoener (1997) Adaptive differentiation following experimental island colonization in Anolis lizards. *Nature* **387**: 70–73.

Losos, J.B., T.R. Jackman, A. Larson, K. de Queiroz, and L. Rodríguez-Schettino (1998) Historical Contingency and determinism in replicated adaptive radiations of island lizards. *Science* **279**: 2115–2118.

Luria, D., (1852) כתבי הגאון ר' דוד לוריה, הפירוש על פרקי דרבי אליעזר Jerusalem: Zonnenfeld. (1990 New revised edition).

Makalowski, Wojciech (2003) Not Junk After All. *Science* **300(5623)**: 1246–1247.

Malthus, Thomas Robert (1798) *An Essay on the Principle of Population* New York: Oxford University Press (1999) Text of original publication available on line at http://files.libertyfund.org/files/311/Malthus_0195_EBk_v6.0.pdf

Mattick, J. S. (2009) Deconstructing the dogma: A new view of the evolution and genetic programming of complex organisms. *Annals of the New York Academy of Sciences* **1178**: 29–46.

Maynard-Smith, J. (1966) Sympatric speciation. *American Naturalist* **100**: 637–650.

Maynard Smith, John (1993) The Theory of Evolution Cambridge: Cambridge University Press (Canto edition).

Mayr, E. (1963) *Animal Species and Evolution* Belknap Press, Harvard University.

Mayr, E. (2004) 80 Years of Watching the Evolutionary Scenery. *Science* **305**: 46–47.

McCarthy et al. (2012) Financial Costs of Meeting Global Biodiversity Conservation Targets: Current Spending and Unmet Needs. *Science* **338 (6109)**: 946–949.

McClintock, B. (1941) Spontaneous alterations in chromosome size and form in *Zea mays*. *Cold Spring Harbor Symposium on Quantitative Biology* **9**: 72–80.

McClintock, B. (1950) The origin and behavior of mutable loci in maize. *Proceedings National Academy of Sciences, USA* **36**: 344–355.

McClintock, B. (1955) Intranuclear systems controlling gene action and mutation, *Brookhaven Symnposia in Biology* No. 8, Mutation, pp. 58–74.

McClintock, B. (1956) Controlling elements in the gene. *Cold Spring Harbor Symposium Quantitative Biology* **21**: 197–216.

McClintock, B. (1983) The significance of responses of the genome to challenge, Nobel lecture, 8 December, 1983, reprinted in *Science* **226**: 792–801 (1984).
McClintock, B (1984) The significance of responses of the genome to challenge. *Science* **226**: 792–801.
McKean, David et al. (1984) Generation of antibody diversity in the immune response of BALB/c mice to influenza virus hemagglutinin. *Proceedings of the National Academy of Sciences USA* **81**: 3180–3184.
McPheron, B. A., D. C. Smith, and S. H. Berlocher (1988) Genetic differences between Rhagoletis pomonella host races. *Nature* **336**: 64–66.
Meselsohn, M. and F. W. Stahl (1958) The replication of DNA in *Escherichia Coli*. *Proceedings of the National Academy of Science. USA* **44(7)**: 671–682.
Messier, François (1985) Social organization, spatial distribution, and population density of wolves in relation to moose density. *Canadian Journal of Zoology* **63(5)**: 1068–1077.
Messier, François (1994) Ungulate population models with predation: A case study with the North American moose. *Ecology* **75(2)**: 478–488.
Messier, François and Michel Crête (1985) Moose-wolf dynamics and the natural regulation of moose populations. *Oecologia* **65**: 503–512.
Messier, François and Damien O. Joly (2000) Comment: Regulation of moose populations by wolf predation. *Canadian Journal of Zoology* **78(3)**: 506–510.
Miller, S. L., (1953) A production of amino acids under possible primitive earth conditions. *Science* **117**: 528–529.
Montealegre-Z. Fernando et al. (2012) Convergent Evolution Between Insect and Mammalian Audition. *Science* **338**: 968–971.
Moorhead, P. S. and M. M. Kaplan (eds.), (1967) *The Mathematical Challenges to the NeoDarwinian Interpretation of Evolution*, Philadelphia: Wistar Inst. of Anat. and Biol.
Morris, Simon Conway (2009) The predictability of evolution: glimpses into a post-Darwinian world. *Naturwissenschaften* **96**: 1313–1337.
Mukerji, M. and S. Mahadevan (1997) Cryptic genes: evolutionary puzzles.(Review article) *Journal of Genetics Indian Academy of Sciences* **76(2)**: 147–159.
Nagel, Thomas (2012) Mind And Cosmos: Why the Materialist Neo-Darwinian Conception of Nature Is Almost Certainly False. Oxford University Press.
NAS (2008) Science, Evolution, and Creationism Committee on Revising Science and Creationism: A View from the National Academy of Sciences, National Academy of Sciences and Institute of Medicine of the National Academies
National Park Service Website, Devil's Hole, updated as of 10 October 2012 http://www.nps.gov/deva/naturescience/devils-hole.htm
National Wildlife Refuge System website updated as of November 7, 2012 http://www.fws.gov/refuges/whm/endangered.html
Nature editorial (2007) Evolution and the brain, *Editorial Nature* **447**: 753 (June 14).
Neuweiler G. (2003) Evolutionary aspects of bat echolocation. *Journal of Comparative Physiology* **A 189**: 245–256.

Nichols R (2001) Gene trees and species trees are not the same. *Trends in Ecology and Evolution* **16**: 358–364.

Niedźwiedzki, Grzegorz, Piotr Szrek, Katarzyna Narkiewicz, Marek Narkiewicz and Per E. Ahlberg (2010) Tetrapod trackways from the early Middle Devonian period of Poland. *Nature* **463**: 43–48.

Ohno, Susumu (1972) So Much "Junk DNA" in our Genome. *Brookhaven Symposium on Biology: Evolution in Genetic Systems* **23**: 366–370.

Olson, E. C. (1965) Summary and Comment. *Systematic Zoology* **14 (4)**: 337–342.

Oparin, A. I. (1953) *The Origin of Life* (Translated from the Russian by S. Margulis) Dover, New York. (Originally published 1938 by Macmillan Co. First Russian edition published in 1924, republished in 1936).

Orgel, L.E. and Crick, F.H. (1980) Selfish DNA: The ultimate parasite. *Nature* **284**: 604–607.

Orgel, Leslie, (1998) The origin of life — a review of facts and speculations. *Trends in Biochemical Sciences* **23**: 491–495.

Orr, H. Allen (1991) A test of Fisher's theory of dominance. *Proceedings of the National Academy of Science. USA* **88**: 11413–11415.

Parker, L, L., P. W. Betts, and B. G. Hall (1988) Activation of a Cryptic Gene by Excision of a DNA Fragment. *Journal of Bacteriology* **170**: 218–222.

Parker, L. L. and B. G. Hall (1990) Mechanisms of Activation of the Cryptic cel Operon of *Escherichia coli* K1. *Genetics* **124**: 473–482.

Pauli, W. (1954) Naturwissenschaftliche und Erkenntnistheoretische Aspekte der Ideen vom Unbewussten. *Dialectica* **8(4)**: 283–301.

Pearson, D. L., Stemberger, S. L. (1980) Competition, body size and the relative energy balance of adult tiger beetles (*Coleoptera: Cicindelidae*). *American Midland Naturalist* **104**: 373–377.

Pérez, Julio E., Carmen Alfonsi, Mauro Nirchio and Sinatra K. Salazar (2008) Bioinvaders: The acquisition of new genetic variation. *Interciencia* **33(12)**: 935–940.

Pigliucci, Massimo (2010) Nonsense on Stilts: How to Tell Science From Bunk. Chicago: University of Chicago Press.

Pimm, S. L., (1988) Rapid morphological change in an introduced bird. *Trends in Evolution and Ecology* **3**: 290–291.

Prevosti, A. et al. (1988) Colonization of America by *Drosophila subobscura*: experiment in natural populations that supports the adaptive role of chromosomal-inversion polymorphism. *Proceedings of the National Academy of Sciences USA* **85**: 5597–5600.

Prijambada, I.D., Seiji Negoro, Tetsuya Yomo, and Itaru Urabe (1995) Emergence of Nylon Oligomer Degradation Enzymes in *Pseudomonas aeruginosa* PAO through Experimental Evolution. *Applied and Environmental Microbiology* **61(5)**: 2020–2022.

Princeton University (2012, October 25) Far from random, evolution follows a predictable genetic pattern. ScienceDaily. Retrieved October 25, 2012, from http://www.sciencedaily.com /releases/2012/10 /121025130922.htm. (See Zhen et al. 2012).

Reynolds, A. E., S. Mahadevan, S. LeGrice, and A. Wright. (1986) Enhancement of bacterial gene expression by insertion elements or mutations in a CAP-cAMP binding site. *Journal of Molecular Biology* **191**:85–95.

Reznick, D. A. and H. Bryga, (1987) Life-history evolution in guppies (*Poecilia reticulata*): 1. Phenotypic and genetic changes in an introduction experiment. *Evolution* **41**: 1370–1385.

Reznick, D. A. and H. Bryga, (1996) Life-history evolution in guppies (*Poecilia reticulata: Poeciliidae*): IV. Parallelism in life-history phenotypes. *American Naturalist* **147**: 319–338.

Reznick, D. A., H. Bryga, and J. A. Endler (1990) Experimentally induced life-history evolution in a natural population. *Nature* **346**: 357–359.

Rogers, A. R. (2011) The Evidence for Evolution. Chicago: University of Chicago Press.

Rosenberg, Noah A. and James H. Degnan (2010) Coalescent histories for discordant gene trees and species trees. *Theoretical Population Biology* **77**: 145–151.

Rutherford S. L. and Lindquist S. (1998) Hsp90 as a capacitor for morphological evolution. *Nature* **396**: 336–342.

Sacks, Jonathan and Richard Dawkins (2012) Debate on Science and Religion http://www.youtube.com/watch?v=roFdPHdhgKQ

Salvini-Plawen, L. V. and E. Mayr (1977) On the evolution of photoreceptors and eyes. In Hecht, Steare, and Wallace (eds.) *Evolutionary Biology* **10**: 207–263.

Sato, A., H. Tichy, C. O'hUigins, P. R. Grant, B. R. Grant, and J. Klein (2001) On the origin of Darwin's Finches. *Molecular Biology and Evolution* **18**: 299–311.

Savolainen, Vincent, Marie-Charlotte Anstett, Christian Lexer, Ian Hutton, James J. Clarkson, Maria V. Norup, Martyn P. Powell, David Springate, Nicolas Salamin and William J. Baker (2006) Sympatric speciation in palms on an oceanic island. *Nature* **441**: 210–213.

Schneider, Tom (2000) Evolution of Biological Information. *Nucleic Acids Research* **28**: 2794–2799.

Schnetz, K. (1995) Silencing of *Escherichia coli* bgl promoter by flanking sequence elements. *EMBO J.* **14**: 2545–2550.

Schnetz, K. and B. Rak (1988) Regulation of the bgl operon of *Escherichia coli* by transcriptional anti-termination. *EMBO J.* **7**: 3271–3277.

Schnetz, K. and B. Rak (1992) IS5: a mobile enhancer of transcription in *Escherichia coli*. *Proceedings of the National Academy of Sciences USA* **89**: 1244–1248.

Schoener, T. W. (1970) Size patterns in West Indian Anolis lizards. II. Correlations with the sizes of particular sympatric species — displacement and convergence. *American Naturalist* **104**: 155–174.

Schoener, T. W. and A. Schoener (1983) The time to extinction of a colonizing propagule of lizards increases with island area. *Nature* **302**: 332–334.

Seong, Ki-Hyeon, Dong Li, Hideyuki Shimizu, Ryoichi Nakamu and Shunsuke Ishii (2011) Inheritance of Stress-Induced, ATF-2-Dependent Epigenetic Change. *Cell* **145**: 1049–1061.

Shapiro, James A. and Richard von Sternberg (2005) Why repetitive DNA is essential to genome function. *Biological Reviews of the Cambridge Philosophical Society* **80:** 1–24.

Shapiro, J. A. (1984) Observations on the formation of clones containing araB-lacZ cistron fusions. *Molecular and General Genetics* **194**: 79–90.

Shapiro, J. A. (1992) Natural genetic engineering in evolution. *Genetica* **86**: 99–111.

Shapiro, J. A. (1997) Genome organization, matural genetic engineering, and adaptive mutation. *Trends in Genetics* **13**: 98–104.

Shapiro, J. A. (1997) A Third Way. *Boston Review* February/March.

Shapiro, James A. (1999c) Part I. Introduction *Annals of the New York Academy of Sciences* **870**: 22.

Shapiro, J. A. (2002) Genome organization and reorganization in evolution: Formatting for computation and function. In: From Epigenesis to Epigenetics: The Genome in Context. *Annals of the New York Academy of Sciences* **981**: 111–134.

Shapiro, James A. (2009) Revisiting the Central Dogma in the 21st. *Annals of the New York Academy of Sciences* **1178**: 6–28.

Shapiro, James A. (2009) Letting *Escherichia coli* Teach Me About Genome Engineering. *Genetics* **183**: 1205–1214.

Shapiro, James A. (2011) *Evolution*: A View from the 21st Century, FT Press Science, Upper Saddle River, NJ.

Schilthuizen, Menno (2001) *Frogs, Flies, and Dandelions*. New York: Oxford University Press.

Simpson, G. G. (1960) The world into which Darwin led us. *Science* **131**: 966–974.

Slack A., Thornton P. C., Magner D.B., Rosenberg S. M., Hastings P. J., (2006) On the Mechanism of Gene Amplification Induced under Stress in *Escherichia coli*. *PLoS Genetics* **2(4)**: 0385–0398.

Smit, A.F.A. (1999) Interspersed repeats and other momentos of transposable elements in mammalian genomes. *Current Opinion in Genetics and Development* **9**: 657–663.

Smith, D. C. (1988) Heritable divergence of *Rhagoletis pomonella* host races by seasonal asynchrony. *Nature* **336**: 66–67.

Snaydon, R. W. and M. S. Davies (1972) Rapid population differentiation in a mosaic environment, II. Morphologicical variation in *Anthoxanthum odoratum*. *Evolution* **26**: 390–405.

Spetner, L. M. (1964) Natural selection: an information-transmission mechanism for evolution. *Journal of Theoretical Biology* **7**: 412419.

Spetner, L. M. (1966) Mutation — the pacemaker of evolution. *Proceedings 2nd International Congress on Biophysics*, Vienna.
Spetner, L. M. (1968) Information transmission in evolution. *IEEE Transactions on Information Theory* **IT14**: 36.
Spetner. L. M. (1970) Natural selection versus gene uniqueness. *Nature* **226**: 948949.
Spetner. L. M. (1997) *Not by chance! Shattering the Modern Theory of Evolution.* Brooklyn: Judaica Press.
Spuhler, J. N. (1985) Anthropology, evolution, and "Scientific Creationism." *Annual Review of Anthropology* **14**: 103–133.
Stearns, Stephen C. (1983) The evolution of life-history traits in mosquitofish since their introduction to Hawaii in 1905: Rates of evolution, heritabilities, and developmental plasticity. *American Zoologist* **23(1)**: 65–75.
Steen, R.G. (2007) *The Evolving Brain: The Known and the Unknown.* Amherst: Prometheus Books
Strick, James (1999) Darwinism and the origin of life: The role of H.C. Bastian in the British spontaneous generation debates, 1868–1873. *Journal of the History of Biology* **32**: 51–92.
Swift, H. (1950) The constancy of desoxyribose nucleic acid in plant nuclei. *Proceedings of the National Academy of Sciences of the USA* **36**: 643–654.
Syvanen, Michael (2012) Evolutionary Implications of Horizontal Gene Transfer. *Annual Review of Genetics* **46**: 339–356.
Thoday, J. M., Gibson, J. B. (1962) Isolation by disruptive selection. *Nature* **193**: 1164–1166.
Thomas, C. A. (1971) The Genetic Organization of Chromosomes. *Annual Review of Genetics* **5**: 237–256.
Thompson, J. N., 1998. Rapid evolution as an ecological process. *Trends in Ecology and Evolution* **13**: 329–332.
Thorne N., C. Chromey, S. Bray, and H. Amrein (2004) Taste perception and coding in Drosophila. *Current Biology* **14**: 1065–1079.
Travisano, M., J. A. Mongold, A. F. Bennett, and R. E. Lenski, (1995) Experimental tests of the roles of adaptation, chance, and history in evolution. *Science* **267**: 87–90.
Tsuchiya, Kozo, Shiro Fukuyama, Naoyuki Kanzaki, Kazuo Kanagawa, Seiji Negoro, and Hirosuke Okada (1989) High homology between 6-aminohexanoate-6-Aminohexanoate-dimer hydrolase of cyclic-dimer hydrolases of Flavobacterium and Pseudomonas strains. *Journal of Bacteriology* **171**: 3187–3191.
Via, Sara (2001) Sympatric speciation in animals: the ugly duckling grows up. *Trends in Ecology and Evolution* **16 (7)**: 381–390.
Waddington, C. H. (1942) Canalization of development and the inheritance of acquired characters. *Nature* **150**: 563–565.

Wanner, B. L., (1985) Phase Mutants: Evidence of a physiologically regulated 'change-in-state' gene system in *Escherichia coli*, in Simon, M. and I. Herskowitz (eds.), Genome Rearrangement. *Proceedings of the UCLA Symposium Apr 7–13 1984*, New York: Alan R. Liss, pp. 103–122.

Warrick D. R., B. W. Tobalske and D. R. Powers. 2005. Aerodynamics of the hovering hummingbird. *Nature* **435**:1094–1097.

Watson, J. D. and F. H. C. Crick, (1953) Molecular structure of nucleic acids. A structure for deoxyribose nucleic acid. *Nature* **171**: 737738.

Welch, K. C., B. H. Bakken, C. M. del Rio, and R. K. Suarez (2006) Hummingbirds fuel hovering flight with newly ingested sugar. *Physiological and Biochemical Zoology* **79**:1082–1087.

Wells, Gary L. and Elizabeth A. Olson (2003) Eyewitness Testimony. *Annual Review of Psychology* **54**: 277–295.

Wen, Yan-Zi, Ling-Ling Zheng, Liang-Hu Qu, Francisco J. Ayala and Zhao-Rong Lun (2012) Pseudogenes are not pseudo any more. *RNA Biology* **9(1)**: 27–32.

West-Eberhard, M. J., (1989) Phenotypic plasticity and the origins of diversity. *Annual Review of Ecological Systems* **20**: 249–278.

West-Eberhard, Mary Jane (2005) Developmental plasticity and the origin of species differences. *Proceedings of the National Academy of Sciences USA* **102** (suppl. 1): 6543–6549.

Wiedersheim, R. 1895 The Structure of Man: An Index to His Past History. Translated by H. and M. Bernard. London: Macmillan and Co.

Wigglesworth, V. B., (1961) Insect polymorphism — a tentative synthesis. *Royal Entomological Society of London. Symp. No. 1: Insect Polymorphism.* pp. 103–113 (as cited by West-Eberhard 1989).

Wilf, Herbert S. and Ewens, Warren J. (2010) There's plenty of time for evolution. *Proceedings of the National Academy of Sciences USA* **107(52)**: 22454–22456.

Wood T. E., J. M. Burke and L. H. Rieseberg (2005) Parallel genotypic adaptation: When evolution repeats itself. *Genetica* **123**: 157–170.

Wright, Barbara E. (2000) A Biochemical Mechanism for Nonrandom Mutations and Evolution. *Journal of Bacteriology* **182**: 2993–3001.

Wu, P., T.-X. Jiang, S. Suksaweang, R. B. Widelitz and C.-M. Chuong (2004) Molecular Shaping of the Beak. *Science* **305(5689)**: 1465–1466.

Wynne-Edwards, V. C., (1986) *Evolution Through Group Selection*. London: Blackwell.

Yockey, H.P. (2005) *Information Theory, Evolution, and The Origin of Life*. Cambridge University Press.

Zhang, Jianzhi (2003) Evolution by gene duplication: an update. *TRENDS in Ecology and Evolution* **18**: 292–298.

Zhen, Ying, Matthew L. Aardema, Edgar M. Medina, Molly Schumer and Peter Andolfatto (2012) Parallel Molecular Evolution in an Herbivore Community. *Science* **337**: 1634–1637.

Index

A

abiogenesis • 15, 17–21, 20n7, 28. *See also* vitalism
Abuha, Rabbi • 101
Abzhanov, A. • 76
Acanthostega fossil • 102–103
acetohydroxyacid synthetase • 53
ACTH • 58
adaptive mutations • 25, 28–29, 32–33, 35, 36, 37–38, 48, 49, 109–113, 131–132
Agard • 113
Ahituv, N. • 51, 142
allopatric speciation • 73, 80–82
amino acids • 53, 88, 118
amplification, of genes • 47
Andersson, J.O. • 97, 98
Anfinsen, C.B. • 35
antibiotics • 10, 31, 53, 97, 116–120
apes • 10, 21–22, 129, 137, 138
arginine vasotocin (AVT) • 65
atheism • 28, 132, 135, 137
AVT (arginine vasotocin) • 65
Axel, R. • 94
Ayala, F. • 130–131

B

bacteria • 43, 44–45, 52–57, 97, 110–113, 117
Baldwin, J.M. • 60–61
Baldwin effect • 61
Barluenga, M. • 59
Barroso, I. • 123
Barton, N.H. • 59, 70
base pairs • 30, 39, 50–51, 53n6, 140, 140n2
Bastian, H.C. • 19
bats • 92–93, 95

Behe, M. • 8
Bejerano, G. • 51
Bergerud, A.T. • 115
Betts, P.W. • 53
Big Bang • 132n1
biogeographical data • 106–109
birds • 21, 64, 74–76, 77, 95–96, 103–104, 138
Bishop, J.A. • 116
Blass, J. • 69–70
Blind Watchmaker, The (Dawkins) • 7
Blount, Z.D. • 91
Bmp4 (*bone morphogenic protein 4*) • 76
Bollinger, R.R. • 121
Bone • 113
bone morphogenic protein 4 (*Bmp4*) • 76
Bossuyt, F. • 94, 145
bp's. *See* base pairs
Bradshaw, A.D. • 62–63, 115
brain, human • 13–14, 35
Brown, J.K. • 95
Brown, W.L., Jr. • 77
Bryga, H. • 71
Buck, L. • 94
Burns, J. • 138

C

Canfield, D. • 34
Caporale, L.H. • 44
Carroll, R. • 104
Carroll, S.P. • 71, 78
Case, T.J. • 72
Cassone, V. • 137
catecholamines • 58
cells • 7–8, 17–20, 17n3, 18n4, 18n6, 21, 22, 23, 28, 32, 37, 44–46, 47, 48, 52–57, 58, 62, 64, 89, 94, 97, 110, 111–113, 118–119, 126, 129, 140, 140n2

Chain, E. • 116–117
character displacement • 76–78
Chen, Y. • 43
CNG (conserved nongenic sequences) • 50–52
Cody, M. • 68–69
Cohen, P. • 58
Common Descent • 8, 15–16, 20–22, 28–29, 35, 36n10, 37, 40, 68–70, 71, 72, 82–83, 85–88, 93, 99–101, 102–103, 105–106, 113, 117–118, 119–120, 129–132, 139
computers • 127
computer simulations • 29–31
Conant, S. • 75
conserved nongenic sequences (CNG) • 50–52
convergent evolution • 73, 89–98, 104–105, 145
Cook, L.M. • 116
Copernican Revolution • 130
Copernicus, Nicolaus • 130
copying errors • 35, 48, 49, 56, 57, 65–66, 72, 74, 75, 76, 118, 132, 145
Cosmic-Ray Group • 16
Coyne, J.A. • 15, 86, 102, 107–109, 110, 120–122, 125, 132
creation • 14–15, 28, 87–88, 100–102, 107–108, 114, 120, 122, 123, 125, 126, 128, 132–138. *See also* Creationism; theological arguments
Creationism • 69, 107–108, 122, 124. *See also* creation; theological arguments
Crête, M. • 114
Crick, Francis • 14, 35, 39
cryptic genes • 43, 46–47, 52–54, 74
Cullis, C.A. • 43
C-value paradox • 124, 140–143, 140n2, 140n3, 141n4
cyclic dimer • 54–56, 55

D
Daeschler, E.B. • 102
Dalos • 88
Darwin, Charles • 16–18, 16n2, 19, 20n7, 23, 42–43, 43n2, 74–76, 79, 80, 94n, 113–114, 114n10, 130–131, 132, 133–134, 135–136, 138, 144
Darwinian revolution • 130
Darwin's Black Box (Behe) • 8
Davies, M.S. • 78
Dawkins, Richard • 7, 15, 30, 86, 87–89, 100, 102, 105, 107–109, 114–115, 122, 124, 132, 136–137
D'Costa, V.M. • 117
Degnan, J.H. • 88
de Lamarck, Jean Baptiste • 38, 42–43, 43n2, 60
deoxyribonucleic acid. *See* DNA
De revolutionibus orbium celestium (Copernicus) • 130
Dermitzakis, E.T. • 50
design. *See* poor design, argument from
Diamond, J.M. • 69
Dickinson, M • 95
Dieckmann, U. • 59
DNA • 18, 18n4, 18n5, 22, 35, 39–40, 43–44, 46, 48–49, 118, 119, 121–125, 140–143, 142
Dobler, S. • 96
Dobzhansky, T. • 38, 79
Doebeli, M. • 59
Doolittle, W.F. • 124, 142
Drake, J.W. • 35, 39
Drosophila • 38, 43, 53, 94
Drosophila subobscura • 43
Drummond, A.J. • 88
Duerden, J.E. • 61
Dunn, E.R. • 79
Dunn, R. • 121
DuPont • 54

E
Eberhardt, L.L. • 114
echolocation • 92–93
Ecker, J.R. • 123

E. coli • 44, 47, 53, 110–111
Eddy, S.R. • 142
Eden, M. • 25
Eldredge, N. • 80–82, 99
Emera, Deen • 98
endangered species • 143–144
environmental inputs • 42–48, 51–52, 60–61, 63–65, 71–72, 73–74, 74–76, 90, 90n1, 103, 106–109, 139, 143, 144, 145
enzymes • 44–45, 53, 54–57, 58, 110–113, 118
epigenetic events, nonrandom • 44–49
epigenetics • 41–49, 53, 57
Erwin, D.H. • 79
eukaryotes • 39, 97–98
Eusthenopteron fossil • 102–103
ev (computer program) • 29
evolution. *See* Common Descent; neo-Darwinian theory (NDT)
Ewens, W.J. • 29
eyes, evolution of • 16–17, 94, 94n2, 125, 125n11, 126, 126n12, 128

F
Feder, J.L. • 59, 70
Fenchel, T. • 77
Fernald, R.D. • 94
Filchak, K.E. • 71
Filipchenko, Y. • 79
finches • 64, 74–76, 77, 103–104, 138
first messengers • 58, 64, 139
fish • 58, 64–66, 71–72, 78, 92–93, 102, 143
Fisher, J.A. • 117
Fisher, R.A. • 38, 63
Flavobacterium • 54–57
Fleischer, R.C. • 95, 145
Fleming, A. • 116
fossil record • 9–10, 16, 16n2, 21–22, 28, 80–82, 99–106
fossils, intermediate • 22, 80, 102. *See also* fossil record

fossils, transitional • 16n2, 102, 102n6. *See also* fossil record
founder effect • 81
Franze, K. • 126
frogs • 94, 141
Fry, B.G. • 95
Futuyma, D.J. • 129

G
Galápagos Islands • 64, 74–76, 77
Galilei, Galileo • 130
gametes • 32, 35, 47, 62, 140, 140n2
Gartner, T.K. • 119
gene duplication • 33, 37
gene pool • 15
genes • 15, 33, 37, 39, 43, 44–45, 46–47, 48, 50–54, 52n5, 56–57, 74, 97–99, 98n4, 110–112, 113, 117, 120–125
genes, cryptic • 43, 46–47, 52–54, 74
genetic changes, nonrandom • 44, 48
genetic information • 18–19, 23, 32, 100
genetic rearrangements, stress as trigger of • 46–48, 53, 56, 57, 62, 63–64, 65, 66, 139, 143
genetic variety • 32–33, 40
genome • 18, 37, 50–52
genome rearrangement • 48
genomes, point mutations and • 44
genotypes • 32, 46, 62, 78
geographical distribution • 106–109. *See also* environmental inputs
geological strata • 99–101, 103–104
Gibson, J.B. • 59
Giovannoni, S.J. • 37
giraffes • 38, 126
Glass, B. • 7
global cooling • 143
Goldberg, S. • 91–92
Gordon, S.P. • 71
Gottlieb, G. • 44
Gould, Stephen Jay • 21, 80–82, 90–91, 99, 129
Grant, P.R. • 75, 77

Grant, R. • 77
Gregory, T.R. • 142

H
Hall, B.G. • 46, 47, 52, 53, 110–111
Heled, J. • 88
heritable changes • 15, 23, 31–32, 36, 38, 41–43, 43n2, 45, 47, 60–64, 68, 70, 72, 78, 107, 110, 139–140
Hersh, M.N. • 47
HGT (horizontal gene transfer) • 97–98, 117
higher organisms, stress on • 58
Hillenmeyer, M.E. • 51
Himmelfarb, Gertrude • 99–100
Ho, M.W. • 44
Hölldobler, B. • 82
horizontal gene transfer (HGT) • 97–98, 117
Hsp90 • 53
Hull, D.L. • 79
human brain • 13–14, 35
human genome • 37, 39, 48, 48n, 122–125, 140n2. *See also* C-value paradox
Huxley, Thomas • 19, 134–135

I
Ichthyostega fossil • 102–103
inducer • 52, 52n5
industrial melanism, evolution of • 115–116
information • 18–19, 23, 28, 37, 71, 74, 100, 105–106, 109–110, 112, 119–120
insects • 31, 70, 77, 89–90, 94–97, 115–116
insertion sequences (IS) • 53–54
intermediate fossils • 22, 80, 102. *See also* fossil record
IS (insertion sequences) • 53–54
Iwama, G.K. • 58

J
Jablonka, E. • 41
Jackman, T.R. • 74
Jacob, F. • 34–35, 44
Jeffroy, O. • 88
Johnson, K.P. • 96
Johnston, T.D. • 44
Joly, D.O. • 114
Journal of Theoretical Biology, The • 7
jumping genes. *See* transposons
"junk DNA" • 39–40, 121–125, 141–142. *See also* DNA

K
Kabbalistic texts • 101
Kakudo, S. • 56
Kanagawa, K. • 56
Keating, J. • 92
Keeling, P.J. • 98
Kepler, Johannes • 130
Kettlewell, H.B.D. • 31, 116
Kinoshita, S. • 54, 56
Kodric-Brown, A. • 95

L
Labin, A.M. • 126
laryngeal nerve • 125, 126–128
latent genes. *See* cryptic genes
Laurin, M. • 121
Lee, C.E. • 44
Lema, S. • 64, 65, 143
leucine • 53
Levy, A. • 99
Levy, S.B. • 117
Lewontin, R.C. • 21, 129
life, origin of • 132–136
Lifshitz, Rabbi Israel • 100–101
Lindquist, S. • 53
linear dimer • 55, 55–56
Linneus, Carolus • 87
Liu, Y. • 93, 145
lizards • 72–74, 77
Logan, Alton • 26–27
Losos, J.B. • 72, 73, 74, 77, 91
Luria, Rabbi David • 108–109

M

macroevolution • 78–81, 129
Mahadevan, S. • 53
Makalowski, W. • 39
Malthus, Thomas • 114, 114n10
matter, living versus nonliving • 18. *See also* abiogenesis; information
Mattick, J.S. • 44
Max, E. • 30n8
Maxwell, James Clerk • 9
Mayr, Ernst • 59, 82, 94
McCarthy, D. • 144
McClintock, Barbara • 46
McKean, D. • 30
McPheron, B.A. • 70, 71
melanism • 115–116
Meselsohn, M. • 35
Messier, F. • 114
Methylophilaceae • 37
microevolution • 78–81. *See also* rapid evolution
Milinkovich, M.C. • 94, 145
Miller, S. • 19
Mishnah • 101n5
mollusks • 143
monkeys • 137. *See also* apes
Monod, J. • 44
monomers • 55, 55–56
Montealegre-Z, F. • 89
Morris, S.C. • 94
Mukerji, M. • 53
mutation rates • 30–31
mutations, health of • 38–39

N

Nagel, T. • 24
National Human Genome Research Institute • 123
natural genetic engineering • 41
naturalism • 132
natural selection • 8–9, 23–24, 31, 35, 47, 47n3, 61–62, 99–100, 113–115, 116, 131–132, 134, 141–142, 144

Nature • 7, 35
N-DNA • 143
NDT (neo-Darwinian theory). *See* neo-Darwinian theory (NDT)
Negoro, S. • 56–57
neo-Darwinian theory (NDT) • 7–11, 23, 42, 44, 47, 63, 65–66, 90–92, 103
neo-Darwinian theory, sympatric speciation and • 58–60
Neuweiler, G. • 93, 145
New Scientist • 132n1
Newton, Isaac • 9, 25, 130, 131
Nichols, R. • 88
Niedźwiedzki, G. • 103
Noble • 44
nonrandom evolutionary hypothesis (NREH) • 10–11, 41–43, 44–49, 58–60, 62, 63–66, 68, 76, 104, 105, 139–140, 142–145
Not By Chance! (Spetner) • 7, 18n5, 36, 41–42, 52n5, 63, 78n3, 112
NREH (nonrandom evolutionary hypothesis). *See* nonrandom evolutionary hypothesis (NREH)
nucleotides • 18, 37, 88
nylA • 56, 57
nylB • 56, 57
nylC • 56
nylon • 54–57

O

Ohno, S. • 39
Olson, E.A. • 27
Olson, E.C. • 79–80
Oparin, Alexander I. • 19
organisms, stress on • 46–48, 53, 56, 57, 62, 63–64, 65, 66, 139, 143
organs, vestigial • 120–123
Orgel, L.E. • 20, 39
Orias, E. • 119
Origin of Species (Darwin) • 19, 130, 133–136, 144
Orr, H.A. • 38

orthogenesis • 79
Overton, J.M. • 68–69

P
paleontology • 99–101
Palmer, J.D. • 98
Panderichthys fossil • 102–103
Parker, L.L. • 52, 53, 91
Pasteur, Louis • 133, 134
Pauli, W. • 24
Pearson, D.L. • 77
peppered moths • 31, 115–116
Peterson, R.O. • 114
phenotypes • 32, 32n9, 38, 42, 46, 49, 62–63, 139
phenotypic plasticity • 62–63, 65
phyla • 50, 78, 95
phylogenetic trees • 86–89, 91–92, 93, 96–100, 139
Pimm, S.L. • 75
plants • 42–43, 46–47, 48n4, 60, 62–63, 67–69, 96, 97, 106–109, 115, 118, 133, 138, 141, 143
plasmids • 53, 53n6, 56, 97
plasticity • 62–63, 65
point mutations • 35–36, 36n10, 38–39, 40, 44, 48, 49, 111, 112, 119
poor design, argument from • 125–128. *See also* creation
population change • 15, 33, 113–115, 117, 129, 131–132
prestin protein • 93
Prevosti, A. • 43
Prijambada, I.D. • 57
probability analysis • 9, 22, 23–25, 28, 29, 36, 36n11, 92, 135–136
Project ENCODE • 123–124, 142
prokaryotes • 97
proteins • 22, 39, 48, 50, 52, 53, 76, 88, 93, 95, 96, 98n4, 113, 118–119, 121, 123
protoplasm • 134–135
pseudogenes • 120–125
Pseudomonas • 56–57

punctuated equilibrium • 80–82
pupfish • 64–65, 143

Q
quantum mechanics • 16

R
Rak, B. • 53
random mutations • 8, 23–24, 25, 28, 29, 31, 32–33, 35, 36, 36n12, 42, 42n1, 44, 47, 48, 50, 60, 61–62, 63, 65–66, 72–73, 76, 77, 78, 80, 81, 86, 92, 97, 103, 105, 107–108, 109–113, 118, 119, 126, 131–132, 141–142, 144
rapid evolution • 44, 68–83
RDH gene • 111–112, 111n7
repressors • 36n10, 52, 52n5, 111–112, 113
reproductive isolation • 70, 71, 81
research • 139–140, 142–143, 145
retina. *See* eyes, evolution of
Reynolds, A.E. • 52
Reznick, D. • 71–72
Ribak, E.N. • 126
ribosomes • 118–119
RNA • 39, 50, 118, 121, 123
Rogers, A.R. • 15, 86, 98–99, 102, 125
Rosenberg, N.A. • 88
Rutherford, S.L. • 53

S
Sacks, Rabbi Jonathan • 124
Salvini-Plawen, L.V. • 94
Sato, A. • 75
Saunders, P.T. • 44
Savolainen, V. • 59
Scaphiopus couchii • 141
Schilthuizen • 59
Schluter, D. • 73
Schneider, T. • 30–31, 36n11
Schnetz, K. • 52–53
Schoener, A. • 72
Schoener, T.W. • 59, 72, 77

second messengers • 58, 64, 139
Shakespeare, William • 70
Shapiro, J.A. • 41, 44, 45, 47, 48, 98–99, 138
Silen • 113
silencers • 52, 53
Simpson, George Gaylord • 133–134, 135, 137
SINE (Short Interspersed Nucleotide Elements) • 48
6-aminohexanoic acid cyclic dimer (Acd) • 54–57, 55
Slack, A. • 47
Smit, A.F.A. • 48
Smith, D.C. • 71
Smith, John Maynard • 59, 134
Snaydon, R.W. • 78
soleucine • 53
somatic cells • 30, 47, 62
species, endangered • 143–144
species selection • 81
Speth • 33
Spetner, L.M. • 8, 29, 30, 36, 37, 44, 73, 90, 117, 119
Stahl, F.W. • 35
Staphylococcus • 117
Stearns, S.C. • 78
Steen, R.G. • 22
streptomycin • 118–119
stress, as trigger of genetic rearrangements • 46–48, 53, 56, 57, 62, 63–64, 65, 66, 139, 143
Strick, J. • 134
Swift, H. • 140n
sympatric speciation • 58–60, 70, 73–74, 81–82
Syvanen, M. • 98

T
Talmud • 101, 108
tetrapods • 102, 103
theological arguments • 88, 100–102, 101n5, 108–109, 114, 120, 125, 128

Thoday, J.M. • 59
Thomas, C.A. • 141
Thompson, J.N. • 78
Thompson, W.R. • 144
Thorne, N. • 94, 145
Tiktaalik fossil • 102–103
toads • 141
Torah • 101
Transactions on Information Theory of the IEEE • 7
transitional fossils • 16n2, 102, 102n6. *See also* fossil record
transposons • 46, 48, 98–99, 98n4. *See also* IS (insertion sequences)
Travisano, M. • 91
Tsuchiya, K. • 56

U
ultraconserved sequences • 50–52, 142
universe, age of • 132, 132n1

V
valine • 53
Variability and Variation (Filipchenko) • 79
vas deferens • 125, 126–127
vermiform appendix • 120–121
vestigial organs • 120–123
Via, S. • 59
vitalism • 17–18, 135
von Sternberg, R. • 48

W
Waddington, C.H. • 41, 60, 61, 62
Wagner, G.P. • 98
Wallace, A.R. • 114n10
Wanner, B.L. • 43
Warrick et al. • 95
Watson, James • 35
Welch et al. • 95, 118–121
Wells, G.L. • 27
Wen, Y. • 122
West-Eberhard, M.J. • 62–63

whales • 92–93
Wiedersheim, R. • 120
Wigglesworth, V.B. • 62–63
Wilf, H.S. • 29
Wilson, Andrew • 27
Wilson, E.O. • 77
Wood, T.E. • 90
Wright, B. • 34
Wu, P. • 76
Wynne-Edwards, V.C. • 115

X
X Club • 134

Y
Yockey, H.P. • 22

Z
Zhang, J. • 33
Zhen, Y. • 96